COLLECTOR'

Rifles

& Muskets

FROM 1450 TO THE PRESENT DAY

COLLECTOR'S GUIDES

Rifles
& Muskets

FROM 1450 TO THE PRESENT DAY

Michael E. Haskew

*To Bruce
Love
Mom & Dad
2014*

amber
BOOKS

Published by
Amber Books Ltd
74–77 White Lion Street
London
N1 9PF
United Kingdom
www.amberbooks.co.uk
Appstore: itunes.com/apps/amberbooksltd
Facebook: www.facebook.com/amberbooks
Twitter: @amberbooks

ISBN 978-1-78274-151-0

Project Editor: Sarah Uttridge
Design: Zoë Mellors
Picture Research: Terry Forshaw

Printed in China

Picture Credits

Accuracy International: 203 top; **Alamy:** 8 (Interfoto), 11 (Bookworm Classics), 19 (J Rowan/Photri Images), 35 (Falkensteinfoto), 53 bottom (Interfoto), 75 (Niday Picture Library), 160 (Zuma Press), 164 (David L Moore – AK), 168 bottom (Interfoto), 171 (Johnny White), 179 top (Daniel Dempster Photography), 181 (Aaron Peterson), 189 (Zuma Press), 200 (Julien McRoberts/Danita Delimont), 202 (Tom Thuleh), 219 (Jeff Morgan); **Art-Tech:** 27, 50, 52, 69, 82, 84, 87, 91 bottom, 93, 119; **Atirador:** 126 bottom; **Benelli:** 162 bottom, 168 top, 186 top, 187, 209; **Beretta:** 178, 179 bottom, 180 top, 185; **Browning:** 160 bottom, 162 top, 177, 184 top, 205 top, 206 both, 216 top; **Teri Bryant:** 73, 113 top; **Bushmaster:** 211; **Cherry's Fine Guns (cherrys.com):** 28 both, 33 top; **Cody Images:** 43 bottom, 48, 60, 98/99, 102, 106, 110/111, 114, 139 bottom, 142 top, 144, 195; **Colt:** 210 top; **Corbis:** 24 (Bettmann), 146 top (David H Wells), 192 (Schultheiss Productions); **CZ:** 208; **Dreamstime:** 21 (Mccool); **Mary Evans Picture Library:** 26 top, 63 (Robert Hunt), 167; **Fabarm:** 176 bottom, 184 bottom; **Fair:** 183 top; **Franchi:** 186 bottom; **Germandaggers.com:** 74 bottom; **Getty Images:** 40 bottom (Archive Photos), 208/209 (AFP); **Heckler & Koch:** 210 bottom; **Henry:** 216 bottom, 217; **William Richard King (kingsforgeandmuzzleloading.com):** 16; **Krieghoff:** 183 bottom; **Library of Congress:** 10, 14, 39 bottom, 44, 54 bottom, 56/57, 70, 89 bottom, 159; **Londonclanger:** 79 (Licensed under the Creative Commons Attribution-share Alike 3.0 Unported Licence); **MacMillan:** 155 top; **Marlin:** 213 both, 214 top; **Mauser:** 191; **Mossberg:** 165, 169 top; **Bertil Olofsson/Krigsarkivet:** 76; **Photos.com:** 23, 36 bottom; **Max Popenker:** 127 top; **Public Domain:** 32, 65; **Remington:** 182, 194 bottom, 207; **Rock Island Auction:** 113 bottom; **Ruger:** 205 bottom; **Salvinelli:** 180 bottom; **Sauer:** 201 bottom; **Tikka:** 203 bottom, 204; **Ukrainian State Archive:** 108, 117; **U.S. Department of Defense:** 6, 12, 29, 122, 125, 128, 130 bottom, 132, 135, 149 bottom, 151, 154; **Winchester:** 156, 190, 212

All profiles © Art-Tech unless credited above

Contents

Introduction

The development, refinement and proliferation of the long arm stretches back more than 600 years. Simply defined as a long barrel, smooth or rifled bore, mounted on a wooden or synthetic stock that fires a projectile, ball, bullet, cartridge or slug, the rifle and shotgun have shaped the course of human history.

LEFT: During field exercises, an infantryman sights a target with his assault rifle, a descendant of the first firearms that date back six centuries in human history.

Since the invention of gunpowder in ninth-century China, innovative minds have contrived means to propel a lethal missile toward a target – man, beast or otherwise. The purpose has been varied, and the long arm has become an indispensible element of civilization as we know it. Early rifles and shotguns were the tools of empire-building and defence, deciding the outcomes of major battles and conflicts between armies East and West.

The long arm has also facilitated the colonization, settlement and economic development of hitherto unexplored, uncharted and little-known areas of the globe. It has been a means of survival, allowing the explorer and the settler to defend, feed and clothe their families. It has allowed generals to conquer. It has been an essential tool, weapon and environmental equalizer throughout modern human history.

As an instrument of warfare the rifle knows no peer in its personal prowess, transforming a man into a soldier. For decades, United States Marine Corps recruits have been required to commit the Rifleman's Creed to memory: 'This is my rifle. There are many like it, but this one is mine. My rifle is my best friend. It is my life. I must master it as I master my life.

My rifle, without me, is useless. Without my rifle, I am useless…'

The evolution of the rifle from the matchlock arquebus, a smoothbore musket widely in use from the fifteenth to the seventeenth century, to the development of the modern assault rifle, automatic rifle and submachine gun, the long arm in its military application has been intended to maximize the combat effectiveness of the soldier. In the civilian world, the rifle has taken its place among hunters and sportsmen, and for many the ownership

BELOW: In this woodcut, an arquebusier stands with his weapon across his shoulder. The arquebus was a muzzleloading firearm that was used from the fifteenth to the seventeenth centuries.

and proper use of the rifle is considered a right of passage for young people. The modern shotgun shares a similar lineage from the seventeenth-century smoothbore blunderbuss that also found applications with the military and in civilian life.

Improving Accuracy and Range

As the name implies, the difference between the rifle and the smoothbore musket is the 'rifling' of the long arm barrel. Rifling is most often defined as the machining of spiral grooves into the barrel of the weapon in order to cause the projectile to spin inside the barrel and during flight as it exits the weapon, resulting in greater accuracy and, in many cases, better range. Although the benefits of rifling had been known for some time, the rifle itself did not come into widespread use until the mid-eighteenth century.

Operational issues with early rifles and the need for a mass-produced long arm, capably filled by existing musket designs, delayed the rifle's introduction and, therefore, its ascendance to primacy among the world's armies and civilian firearm users. The musket, for a time, was sufficient for the battlefield. The musket's ball-shaped projectile was loosely accommodated inside the barrel and tended to bounce off the sides of the barrel when fired. The relatively low muzzle velocity of the projectile resulted in a shorter effective range and a shot that would usually begin to drop precipitously as it neared its target. However, the accuracy of the individual weapon was not as critical when ranks of soldiers were massed together to deliver a powerful volley – literally at times a wall of lead – against an enemy some distance away that was attempting to do the same thing.

Use of the long arm was limited at first due to rapid fouling of the barrel by black powder residue and the heavy smoke the weapon produced when fired, which often obscured targets and diminished any advantage gained by the greater range of the rifle. The time-consuming task of muzzle loading was also a problem. In the military, the early employment of the rifle was almost exclusively by sharpshooters, who most often operated alone. Civilian hunters also began using the rifle.

Although it was determined early in the eighteenth century that an elongated bullet, more aerodynamic than a round lead ball, would travel with greater speed and accuracy, little progress was made in ballistic design until the 1840s when Claude-Étienne Minié (1804–79) and others experimented with a new projectile that would eventually bear his name.

ABOVE: **During the American Civil War, sharpshooters search for targets from the protection of a trench. This eyewitness sketch was completed during a prolonged siege, such as at Petersburg in 1864–65.**

Known as the Minié ball, the projectile revolutionized the employment of the long arm. The conical ball was made of soft lead with external grooves and a hollow, cone-shaped base. When the bullet was fired it expanded, 'grasping' the rifling in the firearm's barrel, closing any existing gap to trap the expanding thrust of the discharging powder and increasing the muzzle velocity of the projectile.

The Minié ball was packaged with the appropriate charge in a paper cartridge that was torn open by the operator. The powder was then poured down the barrel with the bullet following. A ramrod packed the charge tightly, and when the trigger was pulled the process produced the desired improvement in performance. The greater accuracy, velocity and range were accompanied by cleaner firing with reduced fouling of the rifle barrel with powder residue.

Rate of Fire

New rifle designs, such as the breechloader, that offered a much higher rate
of fire than the muzzleloader began to steadily appear during the nineteenth
century. Early rifles sometimes closely resembled the musket in appearance and
operation, and were even referred to as 'rifled muskets'. Improved cartridges
and breechloading mechanisms hastened the adoption of the rifle among
the armies of the world. Conflicts of the mid-nineteenth century, such as the
American Civil War, were marked by extremely high battlefield casualty rates
due to the improvement in rifle technology – the Minié ball had only recently
come into widespread use – that outpaced the refinement of battlefield tactics.
As soldiers in closed ranks blazed away at one another and an accomplished
rifleman was able to load, aim and shoot three rounds a minute, the more
accurate rifle killed and wounded the enemy at a devastating rate.

By World War I, the rifle was the standard-issue long arm of all belligerents. Repeating rifles had been introduced in the later years of the nineteenth century, substantially increasing firepower. Bolt-action rifles that were loaded with multi-round clips produced even greater rates of fire. Submachine guns and automatic weapons carried by individual soldiers soon brought more firepower to bear with an individual soldier than ever believed possible.

Avtomat Kalashnikova 1947

The introductions of the semiautomatic infantry rifle and the world's first operational assault rifles during World War II were followed closely by the advent of the iconic AK-47, the most widely manufactured and distributed infantry weapon of modern times. After more than half a century, the Mikhail Kalashnikov design reigns supreme among military automatic weapons carried by the individual soldier.

The shotgun, meanwhile, developed with the generic term of 'fowling piece', ideal for hunting birds and other small game. Designed as a smoothbore weapon to fire a cartridge of small diameter pellets or 'shot', or to utilize a slug-type cartridge, the shotgun is known for its shorter range and

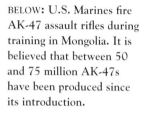

BELOW: U.S. Marines fire AK-47 assault rifles during training in Mongolia. It is believed that between 50 and 75 million AK-47s have been produced since its introduction.

is generally preferred for use with smaller targets that are often moving. In theory, the user of the shotgun may not be required to draw exact aim on a target since the weapon fires a 'spread' of shot that results in a pattern. The expectation is that the pattern of shot will be sufficient and concentrated enough, even without specific aim, to subdue the target.

In addition to hunting small game, the shotgun has historically been used as a defensive weapon by virtue of its supposed ability to stop an adversary. In the American West, guards protecting passengers and valuables on overland stagecoach passages were often armed with shotguns. Early shotguns discharged their loads with a hammer-firing mechanism, and by the mid-nineteenth century the hammerless shotgun, breaking and loading at the breech, was becoming popular and utilized a firing pin to ignite the charge. Later innovations provided for multiple shots as the shotgun was loaded with several rounds in an internal magazine. Lever action, pump action and semiautomatic shotguns were developed and marketed successfully later in the nineteenth century and into the modern era.

Over time, the shotgun has developed for numerous purposes, including hunting, riot or crowd control, and close quarter military engagement. Rifled shotguns firing slug ammunition may be used by the military and for hunting large game, such as deer, with the mass of the slug considered significant enough to inflict a disabling wound.

During World War I, the United States Army introduced the shotgun as an effective weapon for the close fighting that sometimes occurred in the trenches of the Western Front. During World War II, the Vietnam War and later in Iraq and Afghanistan, the shotgun has been a favoured weapon when close proximity to the enemy is a real possibility. Its ability to deliver a disabling shot to an enemy or to afford entry by defeating locks or bolted doors provides an advantage in urban warfare settings. The shotgun is also typically issued to law enforcement officers around the world.

Among sportsmen, the shotgun is employed by those engaged in skeet shooting, trap shooting and sporting clays. These sports enjoy widespread popularity and have long been associated with spirited competition, such as in the Olympic Games.

The evolution of the rifle and shotgun continue to influence society and civilization today. Those who explore their vast history will find an enhanced appreciation and respect for their use as weapons, tools and finely crafted machinery.

Early Rifles & Muskets

The earliest lock, or firing mechanism, for the musket was the matchlock. This innovation allowed a musket-firing soldier, or musketeer, to keep both hands on the weapon and his eye on the target while in combat or training rather than being required to use one hand to ignite the long arm's flash pan with an individually-lit match.

The matchlock first appeared in the mid-1400s, and during the next century it became the primary method of firing the long arm. The matchlock employed a curved lever called a serpentine with a clamp on its upper end to hold a slowly burning match. A lever at the bottom of the musket was connected to the serpentine. As that lever was pulled, the lit match, continually smoldering or burning, was lowered into the flash pan, igniting a priming charge that reached through the touch hole and in turn ignited the main charge to fire the weapon.

By the early 1500s, the wheel-lock mounted on its side a spring-loaded and serrated wheel that rotated itself along with a rotating dog that held a piece

LEFT: The concentrated musket fire of Confederate troops mowed down Union soldiers during ill-advised assaults at Cold Harbor, Virginia, 1864.

MATCHLOCK
COUNTRY OF ORIGIN
Germany
DATE
1450
CALIBRE
10.9mm (.42in)
WEIGHT
4.1kg (9lb)
OVERALL LENGTH
1.2m (48in)
FEED/MAGAZINE
Single shot, muzzleloader
RANGE
45.7m (50yds)

BELOW: The doglock was
an exclusively English
innovation in early firearms
and utilized a flint, frizzen
and safety feature that was
called a 'dog' during the
loading and firing sequence.

of iron pyrites securely in its halves. The musketeer used a spanner to wind
the serrated wheel roughly 75 per cent of its turning radius to set the lock
via a sear mechanism. The dog was subsequently lowered onto the wheel or
a sliding flash pan cover. When the trigger was pulled, the flash pan cover
opened and the iron pyrites dropped onto the wheel. The wheel and pyrites
generated sparks that first ignited the powder in the flash pan and then the
main powder charge that fired the weapon.

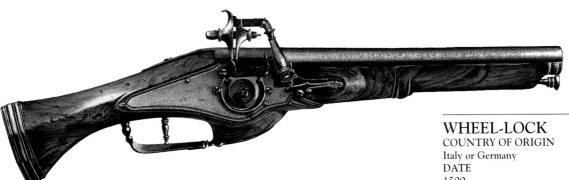

Firing Mechanisms

In the seventeenth century, the English doglock emerged. The forerunner of the flintlock, the doglock included the flint and the frizzen, which was an 'L'-shaped piece of steel covering the flash pan that held the priming charge. It also included a safety known as the 'dog' that engaged when the firing mechanism of the musket was in the half-cocked position. With the doglock, the safety had to be engaged when loading the weapon to prevent premature firing should the cock move forward and strike the frizzen prematurely. When firing the doglock, the cock was moved full forward to allow the dog to fall away to a horizontal position, and the weapon was then fired by pulling the trigger.

The doglock was a uniquely English firearm, and by the dawn of the eighteenth century it had become an emblem of the fighting prowess of the nation's military. While the flintlock was already beginning to emerge, large numbers of British soldiers were undoubtedly armed with the doglock during such significant battles as Blenheim in 1704 and Malplaquet in 1709.

WHEEL-LOCK
COUNTRY OF ORIGIN
Italy or Germany
DATE
1500
CALIBRE
10.9mm (.42in)
WEIGHT
1.81kg (4lb)
OVERALL LENGTH
381mm (15in)
FEED/MAGAZINE
Single shot, muzzleloader
RANGE
27.34m (30yds)

DOGLOCK
COUNTRY OF ORIGIN
England
DATE
1640
CALIBRE
17.53mm (.69in)
WEIGHT
1.81kg (4lb)
OVERALL LENGTH
406.4mm (16in)
FEED/MAGAZINE
Single shot, muzzleloader
RANGE
36.58m (40yds)

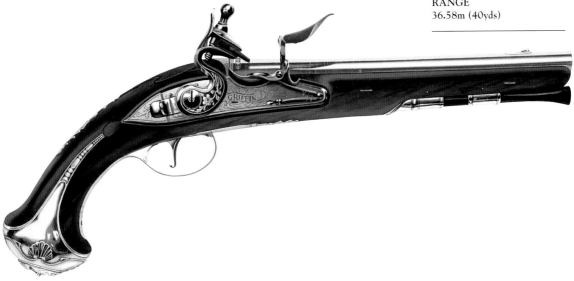

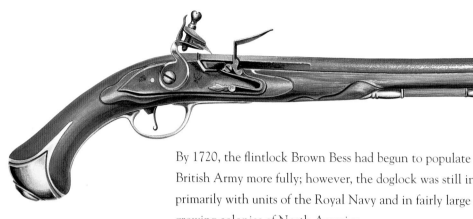

FLINTLOCK
COUNTRY OF ORIGIN
France
DATE
1610
CALIBRE
11.43mm (.45in)
WEIGHT
2.27kg (5lb)
OVERALL LENGTH
406.4mm (16in)
FEED/MAGAZINE
Single shot, muzzleloader
RANGE
50m (54.68yds)

RIGHT: In numerous configurations, the Land Pattern Musket, nicknamed the Brown Bess, was the standard issue long arm of the British Army for more than a century.

By 1720, the flintlock Brown Bess had begun to populate the ranks of the British Army more fully; however, the doglock was still in use, by that time primarily with units of the Royal Navy and in fairly large numbers in the growing colonies of North America.

The primary advantage of the flintlock over the doglock was that its half-cocked position was accomplished with the internal machinery of the musket and the external dog was eliminated. For more than 200 years the flintlock dominated the development of long arms. It was not until the mid-nineteenth century that a substantial number of systems employing percussion caps and then cartridges emerged.

The flintlock itself may be attributed to the ingenuity of French gunsmith Marin le Bourgeoys, who is said to have taken the best attributes of numerous previously developed firing mechanisms and combined them in the first true flintlock, which was presented to King Louis XIII in 1610. Variations on the flintlock theme were developed continually from the 1600s onward, including breechloading versions, rifled versions and more. However, rifling presented difficulties with loading and the vast majority of flintlocks remained smoothbore.

Black Powder Muskets

At the time of the American Revolution, the British Army was equipped with the flintlock Land Pattern Musket, which is known to history as the Brown Bess, a nickname of unknown origin that actually covered a variety of calibres, lengths and lock mechanism variations. The 19mm- (.75in-) calibre Long Land Pattern Brown Bess served as the standard issue long arm of the British Army for more than 100 years, from 1722 to 1838, when it was finally supplanted by a percussion cap musket.

The accuracy of the Brown Bess has been judged as about 160m (175yds). From 1722 to 1768, the Long Land Pattern, with a barrel length of 1168mm (46in) and overall length of 1587mm (62.5in) was in primary service with British infantry. It was supplemented around that time with the Short Land

BROWN BESS
COUNTRY OF ORIGIN
United Kingdom
DATE
1722
CALIBRE
18mm (.71in)
WEIGHT
4.8kg (10.5lb)
OVERALL LENGTH
1490mm (58.5in)
FEED/MAGAZINE
Single shot, muzzleloader
RANGE
91.44m (100yds)

Pattern that had been previously in use with dragoon units. The barrel of the Short Land Pattern measured 1067mm (42in), and its overall length was 1486 mm (58.5in). Smaller in size, the New Land Pattern, New Light Infantry Land Pattern, Cavalry Carbine and Sea Service Pattern were each deployed with service lives of more than 40 years.

With a 990mm (39in) barrel and overall length of 1397mm (55in), the light India Pattern Brown Bess was in general use as an infantry weapon from 1797 to 1854. The India Pattern was so named because a quantity of the muskets was purchased prior to 1797 by the East India Company for use in Egypt.

French Muskets

The French Charleville musket dates to 1717 when it was standardized for the French Army, and it was modified a number of times during the eighteenth century. More than 150,000 were produced, and the 17.5mm- (.69in-) calibre musket remained in service until about 1840. The Charleville musket's barrel was 1130mm (44.5in) and its overall length was 1524mm (60in). One of the Charleville variants, the Model 1777, with a modified stock, cheek rest cut into the butt and a modified trigger guard, was supplied extensively to American soldiers during the Revolutionary War and also carried by French troops who served in North America.

In America, the Model 1795 musket is attributed to inventor Eli Whitney, famous for the cotton gin that contributed to the eventual outbreak of the American Civil War. Whitney based the 17mm- (.69in-) calibre Model 1795 substantially on the French Charleville musket, and it was produced by the Springfield Armoury in Massachusetts just prior to the turn of the nineteenth century. About 150,000 were manufactured, and the service life

CHARLEVILLE
COUNTRY OF ORIGIN
France
DATE
1717
CALIBRE
17.5mm (.69in)
WEIGHT
4.5kg (10lb)
OVERALL LENGTH
1524mm (60in)
FEED/MAGAZINE
Single shot, muzzleloader
RANGE
182.88m (200yds)

was approximately 70 years, encompassing the War of 1812, the War with Mexico and the Civil War. The Model 1795 was also carried by members of the Lewis and Clark expedition to the western reaches of the North American continent. Its effective range was typical of other smoothbore muskets, from 55–69m (60–75yds). Many Model 1795 muskets were later converted from flintlock to percussion.

British military forays into India, Afghanistan and other parts of Central Asia during the 1800s were often opposed by tribesmen armed with the Jezail, a muzzleloading musket that was specifically designed for military use. 'Jezail' is a somewhat generic term that encompasses some long arms that were rifled and manufactured in calibres from 13 to 19mm (.50–.75in). Firing mechanisms ranged from the archaic matchlock to the flintlock that by the mid-nineteenth century was beginning to wane somewhat.

The Jezail was heavy compared to European or American contemporary long arms, but the prowess of the tribesmen during the First Anglo–Afghan War from 1839–42 was acknowledged by the British. One soldier wrote, 'Afghan snipers were expert marksmen and their juzzails [sic] fired roughened bullets, long iron nails or even pebbles over a range of 228 metres [250yds]. The Afghans could fling the large rifles across their shoulders as if they were feathers and spring nimbly from rock to rock…'

Black Powder Rifles

By the time of the American Revolution, Colonel Patrick Ferguson was known far and wide as one of the finest marksmen in the British Army. No doubt that when Ferguson came to the restive colonies in North America he carried one of about 100 experimental flintlocks that he had modified based on an original design developed by Isaac de la Chaumette. The Ferguson rifle, as it was known, was loaded from the top after the

JEZAIL
COUNTRY OF ORIGIN
Afghanistan
DATE
1725
CALIBRE
19.05mm (.75in)
WEIGHT
6.35kg (14lb)
OVERALL LENGTH
1829mm (72in)
FEED/MAGAZINE
Single shot, muzzleloader
RANGE
228.6m (250yds)

FERGUSON
COUNTRY OF ORIGIN
United Kingdom
DATE
1770
CALIBRE
16.51mm (.65in)
WEIGHT
3.5kg (7.5lb)
OVERALL LENGTH
1524mm (60in)
FEED/MAGAZINE
Single shot, breechloader
RANGE
274m (300yds)

soldier made three turns of the trigger guard to open the barrel. Ferguson actually received a patent on his improvements to the original design that dated to 1704 and was probably carrying one of these modified weapons on 7 September 1777 as an officer of the King's 70th Regiment of Foot. Ferguson encountered a tall American officer astride a great chestnut horse along the banks of Brandywine Creek in Pennsylvania.

Although he might easily have taken a fatal shot at the officer, Ferguson later wrote, '…but it was not pleasant to shoot at the back of an unoffending individual, who was acquitting himself very coolly of his duty; so I let him alone.' Later, Ferguson found out that the officer he had come quite close to killing that day was none other than General George Washington, destined to lead the Continental Army to victory in the American Revolution and to become the first President of the United States.

Ferguson's rifle was somewhat ahead of its time, offering a substantially higher rate of fire than other contemporary long arms. However, the Short Land Pattern Brown Bess had recently entered production, and the cost of the Ferguson rifle was prohibitive. It is believed that the Ferguson rifle was used at the Battle of Saratoga in 1777 and during the siege of Charleston in 1780. The rifle was never adopted for mass production due to its cost, production difficulties that emerged from the decentralized manufacturing practices in Britain that were operative at the time and its penchant for breaking down under heavy use.

The Kentucky Rifle
The Kentucky rifle that was prominent during the frontier days of the American settlement west of the Allegheny Mountains is one of a general family of long rifles that were made by American gunsmiths primarily in Kentucky and Pennsylvania beginning around 1700. The barrel of the Kentucky rifle, often 813 to 1219mm (32 to 48in) long, was much longer than its European contemporaries, and the barrel was rifled during a period in which its widespread use remained secondary to the smoothbore musket.

The Kentucky rifle was used both as a hunting rifle and as a military weapon during the French and Indian War, the American Revolution and the War of 1812, as well as during numerous conflicts with Native American tribes during the westward encroachment of settlers in the latter half of the eighteenth century. Often constructed by master gunsmiths carving stocks of beautiful first growth hardwood, the Kentucky rifle sometimes included decorative engraving that richly embellished its lengthy barrel.

In the hands of a skilled marksman, the effective range of the Kentucky rifle was documented in excess of 274m (300yds). One story is told of a shot fired by Daniel Boone during the 1778 siege of Boonesborough, Kentucky, by a combined force of British regulars and their Shawnee allies. An unfortunate officer peered from behind the cover of a large tree and was immediately shot dead by Boone from a distance estimated at 228m (249yds). Approximately 73,000 long rifles were made in America from 1700 to 1900, and their calibre ranged from 6mm (.25in) to 19mm (.75in), and most often was from 10mm (.40in) to 12.19mm (.48in). In 1792, the U.S. Army modified the long rifle and introduced its Model 1792 Contract rifle.

Also known as the Pattern 1800 Infantry Rifle, the Baker rifle was popularly named for its producer, gunsmith Ezekiel Baker of the Whitechapel area of East London. When it was issued in 1800, the Baker rifle was the first standard rifled long arm to enter service with the British Army. The Baker rifle was a muzzleloading flintlock weapon, and the bullet fit so tightly into the grooved barrel that a wooden mallet was sometimes used to complete the loading operation. The 15.9mm- (.625in-) calibre Baker had a barrel length of 769mm (30.3in) and an overall length of 1168mm (46in).

The Baker rifle was in service with elite units of the British Army from 1801 to 1837 and was prominent during the Napoleonic Wars, the War of 1812, in the hands of Americans moving westward during the mid-nineteenth century and during the Texas war for independence from Mexico. Approximately 22,000 Baker rifles were built, and the weapon was known for its accuracy.

KENTUCKY
COUNTRY OF ORIGIN
United States
DATE
1700
CALIBRE
12.19mm (.48in)
WEIGHT
4.54kg (10lb)
OVERALL LENGTH
1778mm (70in)
FEED/MAGAZINE
Single shot, muzzleloader
RANGE
274m (300yds)

Serving under the Duke of Wellington during the Peninsular War, Thomas Plunket of the 1st Battalion, 95th Regiment of Foot, was reported to have shot French General Auguste-Marie-François Colbert at an extended range. Plunket followed that feat of marksmanship by shooting another officer who had come to aid Colbert. A cavalry carbine variant of the standard Baker rifle was also produced.

The Brunswick Rifle

Several rifle designs were considered as replacements for the historic Baker by the mid-nineteenth century, and among the most prominent of these was the Brunswick rifle, a 17.88mm- (.704in-) calibre muzzleloading percussion weapon. The Brunswick rifle was designed in 1836 and produced for the next half century. Early in its evaluation, it was noted that it was heavy and fired a round ball with grooves to take the weapon's rifling. It also could not accommodate the standard paper cartridge then in use with the British military. Nevertheless, the Brunswick was ordered into production with some modification, including a shortened barrel.

In side-by-side testing, the Brunswick was determined to be somewhat more accurate than even the Baker rifle at long ranges. It was durable and required less maintenance than the Baker as well. When it was ordered into production in 1837, the calibre was changed from 16.5mm (.65in) to 17.88mm (.704in). Initial production was distributed among numerous London gunsmiths due to the inability of the Enfield manufacturing facilities to complete the order for 1000 guns in the allotted period of time.

Although it was considered difficult to load, the Brunswick rifle was surprisingly serviceable and equipped British troops during numerous colonial conflicts. During the American Civil War, a quantity of the rifles were imported to the Confederate states and was known to have been employed by Louisiana troops during the siege of Vicksburg.

Surpassed only by the Springfield Model 1861 rifled musket, the British Pattern 1853 Enfield rifled musket was the second most commonly used

---

BAKER
COUNTRY OF ORIGIN
United Kingdom
DATE
1800
CALIBRE
15.9mm (.625in)
WEIGHT
4.1kg (9lb)
OVERALL LENGTH
1168mm (46in)
FEED/MAGAZINE
Single shot, muzzleloader
RANGE
183m (200yds)

LEFT: American frontiersman Daniel Boone sits with his two most trusted companions, his dog and his Kentucky long rifle. In the hands of a trained marksman, the Kentucky rifle displayed remarkable accuracy.

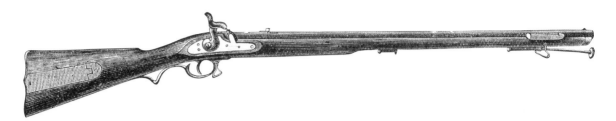

BRUNSWICK
COUNTRY OF ORIGIN
United Kingdom
DATE
1836
CALIBRE
17.88mm (.704in)
WEIGHT
4.5kg (10lb)
OVERALL LENGTH
1160mm (45.67in)
FEED/MAGAZINE
Single shot, muzzleloader
RANGE
274m (300yds)

ENFIELD
COUNTRY OF ORIGIN
United Kingdom
DATE
1853
CALIBRE
14.65mm (.577in)
WEIGHT
4.3kg (9.5lb)
OVERALL LENGTH
1400mm (55in)
FEED/MAGAZINE
Single shot, muzzleloader
RANGE
1800m (1970yds)

infantry long arm of the American Civil War. Estimates of the number of 1853 Enfields imported to America during the war years exceed 900,000. The 1853 Enfield was called a 'rifled musket' simply because its length was equal to that of the older muskets it replaced in the field. It had previously been necessary for the musket barrel to be long enough for the muzzle to protrude past the faces of soldiers in the front rank when firing from behind them in the second rank of a standard formation. Furthermore, the barrel needed sufficient length to effectively deploy a bayonet if necessary.

The 1853 Enfield was a 14.65mm- (.577in-) calibre muzzleloading weapon that fired a variant of the highly effective Minié ball developed by the British in response to the French innovation. The 1853 Enfield was, to a great extent, designed specifically to fire the new ammunition and to keep up with the improving French ballistics. They were made by the Royal Ordnance Factory as well as independently contracted gunsmiths.

The 1853 Enfield employed the percussion lock firing mechanism, successor to the flintlock. Its barrel was 838mm (33in), and the overall length was 1400mm (55in). In the hands of an experienced soldier, its rate of fire was up to three rounds a minute with a maximum range of 1800m (1970yds). More than 1.5 million of the famous weapon were produced from 1853 to 1867, and it remained in service with the British Army until 1889. The 1853 Enfield saw major action during the Crimean War, the Indian Mutiny and the New Zealand Land Wars.

During the early 1850s, the British Army was in the midst of a transition from smoothbore to rifled muskets. With the approval of the 1853 Enfield to enter production by the end of that year, the new long arm was deployed

in great numbers during the Crimean War. The first Enfields to arrive in the Crimea were uncrated and issued to the troops in February 1855.

Three additional American long arms of the early nineteenth century, manufactured at the Harpers Ferry Federal Arsenal in what is now the state of West Virginia, are worthy of note. The flintlock, muzzleloading Harper's Ferry Model 1795 musket fired a 17.5mm (.69in-) calibre ball from a 1130mm (44.5in) barrel. Some gun enthusiasts and scholars differentiate the Harpers Ferry Model 1795 from that produced during approximately the same period at the Springfield Armoury.

The 13.7mm- (.54in-) calibre Harpers Ferry Model 1803 rifle was the first weapon officially placed into production at the behest of the U.S. Army as

ABOVE: The Coldstream Guards concentrated the fire of their Enfield rifles and decimated a Russian charge, sealing the victory for Allied forces at the Battle of the Alma during the Crimean War.

HARPERS FERRY 1795
COUNTRY OF ORIGIN
United States
DATE
1795
CALIBRE
17.5mm (.69in)
WEIGHT
4.54kg (10lb)
OVERALL LENGTH
1524mm (60in)
FEED/MAGAZINE
Single shot, muzzleloader
RANGE
55m (60yds)

HARPERS FERRY 1816
COUNTRY OF ORIGIN
United States
DATE
1816
CALIBRE
17.5mm (.69in)
WEIGHT
4.54kg (10lb)
OVERALL LENGTH
1473mm (58in)
FEED/MAGAZINE
Single shot, muzzleloader
RANGE
91m (100yds)

a standard design. Interestingly, it was not equipped with a bayonet since riflemen were not generally expected to engage in close-quarter fighting with enemy infantry. With minor variations, the 17.5mm- (.69in-) calibre Harpers Ferry Model 1816 musket was manufactured steadily from 1817 to 1844. Many of these flintlock muskets were converted to percussion during the Civil War.

Born in 1766, British gunsmith Joseph Manton is said to have produced some of the finest long arms in the world during the golden age of the flintlock. When Manton was just 29 years old he developed a tool with which the barrel of a musket could be rifled with greater ease. He also influenced the development of modern ammunition by redesigning shot for faster reloading. His tube lock, also known as the pill lock, was an advancement over the flintlock firing system that was never as widely adopted as the percussion lock, although it was placed in service with the Austrian Army.

Percussion Rifles

The successor to the flintlock system, the percussion lock utilized a hammer to strike a small percussion cap that was filled with fulminate of mercury to ignite the firing charge. To fire the percussion lock, the percussion cap was placed over a small nipple that held a tube that entered the barrel. When the hammer struck the percussion cap, a small explosion occurred inside the cap and the flame traveled through the tube to the barrel to ignite the charge.

When Colonel Jefferson Davis, future President of the Confederate States of America, took command of a regiment of U.S. troops from Mississippi during the Mexican War, he was determined that his men should be

equipped with the relatively new Model 1841 rifle. Davis petitioned his
superior, General Winfield Scott, for the rifle and was denied. Undaunted,
he took his case to President James K. Polk, who ruled in Davis' favour. The
Mississippians got their rifles, henceforth to be known as the Model 1841
Mississippi, and the disagreement marked the beginning of a feud between
Davis and Scott that lasted for the rest of their lives.

The Model 1841 Mississippi rifle was manufactured at the Harpers Ferry
arsenal prior to and in the midst of an extensive 1842 refit of the machinery
at the arsenal from the manufacture of smoothbore muskets to the new
weapon. Under the watchful eyes of manager Eli Whitney Blake and
foreman Thomas Warner, who had previous experience at the Springfield
Armoury, the rifle was produced from 1841 to 1861, eventually evolving into
the Model 1855 standard issue rifle and finally the Springfield Model 1861.

The 13.7mm- (.54in-) calibre Mississippi rifle was the first standard issue
U.S. rifle to use the percussion lock. In 1855 the calibre was changed to
14.73mm (.58in) to accommodate the revolutionary Minié ball. 'V' notch

BELOW: American troops
armed with the Model
1841 Mississippi rifle storm
strong enemy positions
during the Mexican War.
Colonel Jefferson Davis
demanded that his troops
be equipped with the
weapon.

M1841 MISSISSIPPI
COUNTRY OF ORIGIN
United States
DATE
1841
CALIBRE
13.72mm (.54in)
WEIGHT
4.2kg (9.25lb)
OVERALL LENGTH
1230mm (48.5in)
FEED/MAGAZINE
Single shot, muzzleloader
RANGE
1006m (1100yds)

SPRINGFIELD 1855
COUNTRY OF ORIGIN
United States
DATE
1855
CALIBRE
14.73mm (.58in)
WEIGHT
4.1kg (9lb)
OVERALL LENGTH
1422mm (56in)
FEED/MAGAZINE
Single shot, muzzleloader
RANGE
270m (300yds)

sights were later replaced by leaf and then ladder sights. The rifle's overall length was 1230mm (48.5in), and its barrel extended to 838mm (33in). Although the Model 1841 Mississippi rifle was considered somewhat outdated by the eve of the Civil War, it was issued to Union and Confederate troops in significant numbers. The smoothbore musket remained in widespread use as well, as many soldiers volunteered for service with whatever shoulder arm was at hand, including old flintlocks and squirrel guns.

The 14.73mm- (.58in-) calibre Springfield Model 1855 rifle may also be considered a rifled musket since its overall length of 1422mm (56in) and barrel length of 1016mm (40in) were equivalent to that of the muskets of the day which it was intended to replace. About 60,000 of the weapons were produced from 1856 to 1860 at the Springfield, Massachusetts, armoury.

The Model 1855 was developed at the time the percussion cap was coming into more common use as a replacement for the outmoded flintlock. However, rather than outfitting the Model 1855 with the standard percussion cap components, its designers decided to use the Maynard tape primer, which was intended to improve the weapon's rate of fire. A tape automatically fed a primer to the proper position every time the hammer was cocked, eliminating the need for the soldier to place an individual percussion cap on the nipple before each shot was fired. The experiment was a failure, as the tapes proved not to be waterproof as anticipated and the feed mechanism often failed. Many soldiers abandoned the Maynard tape in the field and resorted to placing the percussion caps by hand.

The Model 1855 is easily recognizable due to the characteristic hump beneath the hammer that had been necessitated due to the introduction of

the Maynard tape. As the coming of the Civil War loomed, the use of the Maynard tape was discontinued and other alterations were made, resulting in the Springfield Model 1861. The Model 1855 first saw action in September 1858 at the Battle of Four Lakes near present-day Spokane, Washington. It was instrumental in the decisive defeat of a hostile Native American force armed with smoothbore muskets. In the hands of an experienced and well-drilled soldier, the Springfield Model 1855 was capable of firing up to three rounds a minute.

Sir Joseph Whitworth, a prosperous British engineer and inventor, developed the Whitworth rifled musket in the mid-1850s as a potential replacement for the Pattern 1853 Enfield. Whitworth's most obvious departure from the standard with his new rifle was the introduction of a hexagonal barrel, which eliminated the need for grooves in the projectile.

In 1857, the Whitworth bested the 1853 Enfield in a series of side-by-side tests. Most impressive was the Whitworth's ability to consistently hit a target at 1829m (2000yds), fully 549m (600yds) greater distance than the 1853 Enfield. Nevertheless, the design was rejected due to its high cost and propensity to become rapidly fouled with powder residue.

Confederate Sharpshooters
Whitworth did manage to sell some of the 13,000 rifles produced from 1857 to 1865 to the French government. A quantity was also delivered to the Confederacy during the American Civil War. The Whitworth became quite popular among Confederate troops, and competitions were sometimes held to determine who would be issued one of the limited number of Whitworth rifles available. Only the best marksmen received them, and quite a number of Union soldiers came to grief in the sights of a Rebel Whitworth. Most of these rifles had open sights with an adjustable front blade to compensate for wind.

On 9 May 1864, during the Battle of Spotsylvania Courthouse in Virginia, Union General John Sedgwick scolded his staff officers and

WHITWORTH
COUNTRY OF ORIGIN
United Kingdom
DATE
1854
CALIBRE
11.43mm (.45in)
WEIGHT
4.1kg (9lb)
OVERALL LENGTH
1244mm (49in)
FEED/MAGAZINE
Single shot, muzzleloader
RANGE
1400m (1500yds)

RIGHT: British inventor Joseph Whitworth (1803–87) produced the Whitworth rifle, which became famous for its accuracy and is considered by some to be the world's first sniper rifle.

members of a nearby artillery unit for taking cover when they heard the whistling crackle of a bullet that had unmistakably been fired from a Confederate Whitworth rifle. Sedgwick believed the Confederate sharpshooters were as much as 914m (1000yds) distant and boasted, 'What? Men dodging this way for single bullets? What will you do when they open fire along the whole line? I am ashamed of you. They couldn't hit an elephant at this distance.'

Standing in the open seconds later, Sedgwick repeated himself, saying, 'I'm ashamed of you, dodging that way. They couldn't hit an elephant at this distance.' Abruptly, he was struck by a bullet below his left eye and

killed. It was said that five Confederate soldiers claimed to have fired the shot from their trusty Whitworths.

The muzzleloading 11.4mm- (.45in-) calibre Whitworth rifle was 1244mm (49in) long, and its barrel measured 838mm (33in) in length.

American inventor and gunsmith Christian Sharps began work in 1848 on a series of rifles that would bear his name. The best known of the series is the single shot, 13.2mm- (.52in-) calibre military type breechloading falling block percussion rifle that was first used during the Civil War. It was known generally as a Sharps New Model rifle and forever linked to the 2nd United States Sharpshooters commanded by Colonel Hiram Berdan.

The men under Berdan's command dressed in distinctive green uniforms and were required to pass rigorous marksmanship standards. Their light guns weighed only 3.6kg (8lb) and were purchased at an individual cost of $42.50. Firing their rifles at a fantastic rate of eight to 10 rounds a minute, Berdan's Sharpshooters delivered outstanding performance in numerous major battles, including Antietam, Chancellorsville and Gettysburg. A carbine version of the Sharps rifle was considered an excellent cavalry weapon as well until somewhat eclipsed by the introduction of repeating carbines. More than 100,000 Sharps rifles were manufactured from 1850 to 1881, and sporting versions were popular into the 1880s.

A mechanical engineer and inventor, Hiram Berdan had modified many of the Sharps rifles issued to his 2nd U.S. Sharpshooters. In 1868, he developed the Berdan rifle, which became the standard issue long arm of the Imperial Russian Army from 1870 to 1891 when it was replaced by the Mosin-Nagant.

SHARPS
COUNTRY OF ORIGIN
United States
DATE
1848
CALIBRE
13.2mm (.52in)
WEIGHT
4.3kg (9.5lb)
OVERALL LENGTH
1200mm (47in)
FEED/MAGAZINE
Single shot, breechloader
RANGE
460m (500yds)

BERDAN
COUNTRY OF ORIGIN
United States
DATE
1868
CALIBRE
10.6mm (.42in)
WEIGHT
4.2kg (9.3lb)
OVERALL LENGTH
1300mm (51in)
FEED/MAGAZINE
Single shot, breechloader
RANGE
284m (311yds)

Two versions of the Berdan rifle, the 1868 hammerless breechblock design manufactured by Colt in the United States and the 1870 single-shot bolt action model, manufactured at Birmingham Small Arms in Britain and later at several facilities in Russia, were produced for the Russian military. Both were 10.6mm- (.42in-) calibre weapons and were also used extensively in Russia by hunters with sporting and even shotgun variants in production as late as the 1930s. More than three million of all variants were produced, and the rifle was also used by the Finnish, Bulgarian and Serbian military.

Early Cartridge Rifles

Cartridge rifles came into their own and entered widespread use during the mid-nineteenth century. Both European and American designs were prominent in the evolution of the long arm as breechloading rifles eclipsed the older, slower firing muzzleloaders and raised the firepower of land armies to new heights. During the Franco–Prussian War of 1870–71 both the Prussian and French armies employed breechloading, cartridge rifles that became famous.

The primary infantry long arm of the Prussian Army from 1848 until the end of the Franco–Prussian War was the Dreyse Needle rifle, so named because of the thin firing pin that passed through a paper cartridge and struck a percussion cap embedded at the base of the bullet itself. The first rifle to utilize a bolt action to operate the chamber, the Dreyse Needle rifle was invented by German gunsmith Johann Nikolaus von Dreyse and accepted by the Prussians for production in 1841, although it was not placed in service for another seven years.

Employed during the Prussian war with Austria in 1866, the weapon provided a definitive advantage for the Prussian infantry, allowing them to fire up to 12 rounds a minute, a rate several times higher than the Austrian enemy who were still laboring to load and fire outmoded muzzleloaders. For all its innovation, however, the needle gun did have its drawbacks. A significant volume of gas escaped from the chamber when the weapon was fired, reducing muzzle velocity. In turn, the needle gun's effective range was

DREYSE NEEDLE
COUNTRY OF ORIGIN
Prussia
DATE
1841
CALIBRE
15.4mm (.61in)
WEIGHT
4.7kg (10.4lb)
OVERALL LENGTH
1420mm (56in)
FEED/MAGAZINE
Single shot, breechloader
RANGE
600m (650yds)

diminished along with its ability to stop an enemy soldier with its 15.4mm (0.61in) projectile.

Still, this obviously impressive performance of the Dreyse Needle gun validated the French efforts to develop their own breechloading cartridge rifle, the Chassepot. The primary infantry weapon of the French Army during the Franco–Prussian War, the Chassepot was also a bolt action, breechloading weapon. Its introduction in 1866 marked the replacement of an assortment of muzzleloading rifles that continued to fire the venerable Minié ball.

Officially known as the Fusil Modèle 1866, the Chassepot was popularly named for its inventor, Antoine Alphonse Chassepot, who had experimented with numerous breechloading designs prior to his greatest success. The 11mm (0.43in) Chassepot, like the Dreyse, fired a paper cartridge that was ignited by a needle-like firing pin. Its lead bullet was

ABOVE: After seeing the effects of the Prussian Dreyse Needle Gun in combat, the French military recognized the need for such a weapon. The result was the Chassepot, which entered service in 1867.

CHASSEPOT
COUNTRY OF ORIGIN
France
DATE
1866
CALIBRE
11mm (.433in)
WEIGHT
4.6kg (10.4lb)
OVERALL LENGTH
1310mm (51.6in)
FEED/MAGAZINE
Single shot, breechloader
RANGE
1200m (1300yds)

SNIDER-ENFIELD
COUNTRY OF ORIGIN
United Kingdom
DATE
1866
CALIBRE
14.7mm (.577in)
WEIGHT
3.8kg (9lb)
OVERALL LENGTH
1250mm (49.25in)
FEED/MAGAZINE
Single shot, breechloader
RANGE
550m (600yds)

shaped like a rounded cylinder. During the Franco–Prussian War, the Chassepot was proven superior to the Dreyse with much greater range and a rate fire from eight to 15 rounds a minute.

More than a million Chassepot rifles were produced before it was withdrawn from service in 1874 in favour of the Gras rifle, which was quite similar in outward appearance to the Chassepot but utilized a metallic centrefire cartridge. The Chassepot was also purchased by the Tokugawa Shogunate of Japan.

In 1866, the British Army was issued the breechloading Snider-Enfield rifle, a conversion of the Pattern 1853 Enfield muzzleloader via a system devised by American inventor Jacob Snider. Five years later it was supplanted by the Martini-Henry rifle. The 14.7mm- (.577in-) calibre Snider-Enfield was capable of a rate of fire up to 10 rounds a minute, and the conversions were completed at the Royal Small Arms Factory in Enfield with the rifles receiving new breechblocks and receiver assemblies but retaining the original lock, iron barrel, hammer and stock.

Despite its relatively short stint as the primary British Army long arm, the service life of the Snider-Enfield was remarkably lengthy, from 1867 to 1901, and the conversion rifle proved more accurate than the original 1853 Enfield. Numerous variants of the Snider-Enfield were produced, including cavalry, engineer, artillery and yeomanry carbines, as well as a naval rifle. The Snider-Enfield fired the metal Boxer cartridge, named for Colonel Edward Mounier Boxer of the Royal Arsenal, Woolwich, and was the first standard issue British breechloader to fire a metal cartridge. The

rifle was used in action by British troops in Ethiopia and later by the Indian Army until the turn of the twentieth century.

The 11.4mm- (.45in-) calibre Martini-Henry rifle was purpose-built as a breechloader that combined a loading lever and dropping block action with a striker and cocking apparatus enclosed within the receiver, greatly improving the functionality of the weapon. Four models of the Martini-Henry were produced by the Royal Small Arms Factory, and production of the last of these terminated in 1889, although the rifle remained in service in small numbers up to the end of World War I.

Capable of firing up to 12 rounds a minute over sliding ramp rear sights and fixed post front sights, the Martini-Henry had an effective range of 370m (405yds) and a maximum range of 1700m (1859yds). It was present with the 139 soldiers of the 2nd Battalion, 24th Regiment of Foot at Rorke's Drift during the Zulu War of 1879. The small contingent of British soldiers held off thousands of Zulu warriors during repeated assaults. Despite the fact that inferior thin brass cartridges caused the rifles to jam frequently during the conflict, the performance at Rorke's Drift is testament to the battlefield viability of the Martini-Henry.

When the Turkish government was thwarted in its effort to purchase the Martini-Henry from the British because all production was earmarked for the British military, the Rhode Island-based Providence Tool Company sold virtually identical copies to the Turks, and these were used in the Russo–Turkish War of the late 1870s. A shotgun variant, known as the Greener Police Gun, was also produced.

The first standard issue breechloading rifle issued to troops of the United States Army was the Springfield Model 1873 Trapdoor, selected for production after trials of 99 foreign and domestic weapons. Named for the Allin breechblock design that opens like a trapdoor, the Model 1873 was the fifth Springfield variation of that system and became one of the primary weapons of U.S. forces during the prolonged fighting with hostile Native Americans on the western frontier.

MARTINI-HENRY RIFLE 1871
COUNTRY OF ORIGIN
United Kingdom
DATE
1871
CALIBRE
11.4mm (.45in)
WEIGHT
3.83kg (8.7lb)
OVERALL LENGTH
1245mm (49in)
FEED/MAGAZINE
Single shot, breechloader
RANGE
370m (405yds)

SPRINGFIELD TRAPDOOR

COUNTRY OF ORIGIN
United States
DATE
1873
CALIBRE
11.4mm (.45in)
WEIGHT
4kg (8.81lb)
OVERALL LENGTH
1318mm (51.88in)
FEED/MAGAZINE
Single shot, breechloader
RANGE
548.6m (600yds)

RIGHT: When the U.S. 7th Cavalry was wiped out at Little Big Horn, their single-shot rifles were overmatched by the repeating rifles carried by many of the Cheyenne and Sioux warriors they faced.

The 11.4mm- (.45in-) calibre rifle fired a cartridge designated the .45-70-405, denoting the calibre, the 70 grains of black powder propellant and the 405-grain weight of the bullet respectively. The rifle was capable of firing up to 10 rounds a minute, and approximately 700,000 were manufactured at the Springfield Armoury.

At the Battle of Little Big Horn, popularly known as Custer's Last Stand, a detachment of the U.S. 7th Cavalry under General George Armstrong Custer was annihilated by Cheyenne and Sioux warriors. Many of the cavalrymen were armed with the carbine variant of the single shot Model 1873, and in the wake of the massacre concerns about spent cartridges jamming the weapon were raised. The original copper cartridge was subsequently replaced with brass.

Repeating Rifles

While General George Custer and the troopers of the U.S. 7th Cavalry at the Little Big Horn were armed with the single shot Springfield Model 1873 Trapdoor, the Sioux and Cheyenne braves who annihilated them were reportedly armed with the Henry rifle, a lever-action repeating rifle that fired a 11.1mm- (.44in-) calibre rimfire cartridge. At the time of the Battle of Little Big Horn, the Henry rifle was a relatively new innovation, although it had already seen its baptism of fire during the American Civil War.

Designed in 1860 by Benjamin Tyler Henry, the Henry rifle was an improvement over the Volition Repeating rifle designed by Walter Hunt for the Volcanic Repeating Arms Company. Volcanic Repeating Arms survived only from 1855 to 1866; however, its legacy companies, Smith & Wesson and the Winchester Repeating Arms Company, became legendary. The Henry rifle was designed five years after the Volition rifle. It was loaded via a 16-round tube magazine and capable of firing up to 28 rounds a minute.

The Henry was never formally adopted by the U.S. Army; however, some Union soldiers were intent on owning one and paid for their own. By the end of the Civil War, approximately 14,000 had been produced.

Operating the lever with a downward motion ejected the spent cartridge and chambered the next round, which was spring fed from the magazine, while also cocking the hammer. Despite its popularity one criticism of the Henry is that it lacked the muzzle velocity of other period repeating rifles.

One of the chief competitors of the Henry repeating rifle was the Spencer. During the Civil War, the U.S. Army did adopt the Spencer, although it never replaced the standard issue muzzleloading rifles then in use. Designed by American inventor Christopher Spencer, the 13.2mm- (.52in-) calibre Spencer repeating rifle fired up to 20 rounds a minute from a seven-round

HENRY
COUNTRY OF ORIGIN
United States
DATE
1860
CALIBRE
11.18mm (.44in)
WEIGHT
4.5kg (9.9lb)
OVERALL LENGTH
1143mm (45in)
FEED/MAGAZINE
Lever action repeater
RANGE
91m (100yds)

SPENCER
COUNTRY OF ORIGIN
United States
DATE
1860
CALIBRE
13.2mm (.52in)
WEIGHT
4.54kg (10 1lb)
OVERALL LENGTH
1200mm (47in)
FEED/MAGAZINE
Lever action repeater
RANGE
457m (500yds)

tube magazine. More than 200,000 were manufactured from 1860 to 1869, and the Spencer was known to Confederate soldiers as the rifle that could be loaded on Sunday and fired all week. Unlike the Henry, the Spencer hammer was manually cocked. A carbine variant was especially favoured by Union cavalrymen.

In April 1887, the French Army adopted the 8mm (.314in) bolt action Lebel Model 1886 repeating rifle. Firing from an eight-round tube magazine, the Model 1886 was effective to 400m (438yds), and its maximum range was 1800m (1970yds). In 1893, a bolt-action modification was approved and the official name of Fusil Modele 1886 was modified to Fusil Modele 1886 M93.

One of the most numerous rifles produced during the period, nearly
2.9 million of the Lebel 1886 were completed between 1887 and 1920 by
state manufacturers Chatellerault, Saint-Etienne and Tulle. The French
government was motivated by the invention of smokeless gunpowder and
sought a new infantry weapon that could take full advantage of it on the
battlefield, and other European nations followed the trend, adopting small
bore rifles that utilized smokeless powder.

The Lebel M1886 gained a reputation as a rugged and powerful rifle and
served as a stalwart of the French Army during World War I along with
the Berthier rifle, its intended replacement, which entered service in 1907.
In truth, the performance of the Lebel M1886 was rapidly eclipsed by new
models produced in other countries, and by the eve of World War I the
French were experimenting with automatic rifles for potential standard
issue to infantry units. However, the coming of war largely shelved the idea
in favour of the Berthier and the aging Lebel M1886.

The bolt-action Mauser Model 1889 rifle was adopted in Belgium with
modifications to an existing German design. The rifle was developed by
Mauser in the wake of the company's Model 1887 and intended for use
with smokeless powder. A second rifle, the 1891 Argentine, was also
produced for export. The Mauser Model 1889 fired a 7.65mm (.30in)
cartridge fed from a five-round box magazine – the first such rifle in the
Mauser line. While other major European powers declined to adopt the
Model 1889 and the German government rejected the rifle as well, the
Belgians established the famous Fabrique Nationale d'Armes de Guerre, or
FN, to manufacture it.

LEBEL 1886
COUNTRY OF ORIGIN
France
DATE
1886
CALIBRE
8mm (.314in)
WEIGHT
4.41kg (9.7lb)
OVERALL LENGTH
1300mm (52in)
FEED/MAGAZINE
Bolt action repeater
RANGE
400m (438yds)

LEFT: Native Americans
used repeating rifles as
they fought against the
expansion of white settlers.

MAUSER INFANTRY MODEL 1889
COUNTRY OF ORIGIN
Belgium
DATE
1889
CALIBRE
7.65mm (.301in)
WEIGHT
4kg (8.82lb)
OVERALL LENGTH
1295mm (51in)
FEED/MAGAZINE
Bolt action repeater
RANGE
550m (550yds)

MANNLICHER-CARCANO 1891
COUNTRY OF ORIGIN
Italy
DATE
1891
CALIBRE
6.5mm (.26in)
WEIGHT
3.4kg (7.5lb)
OVERALL LENGTH
1015mm (40in)
FEED/MAGAZINE
Bolt action repeater
RANGE
600m (656yds)

RIGHT: The Lee-Metford rifle featured a rear-locking bolt system and was issued to British troops during the Boer War. It was quickly replaced by the Lee-Enfield, although a few were converted to experimental semiautomatic weapons.

Steel Barrel Jacket

One intended innovation with the Model 1889 was a steel jacket that surrounded the barrel, ostensibly to prolong the life of the weapon and reduce wear and tear. In combination with the solid wooden stock, the hope was for a remarkably durable firearm. In the end, the steel jacket proved susceptible to moisture and added little to the longevity of the weapon. However, the Model 1889 served as the primary infantry rifle of the Belgian Army during World War I and remained in service with northern European armies into the 1940s. More than 275,000 were manufactured.

Often known as the Mannlicher-Carcano 1891 for the 6.5mm (.26in) cartridge of the same name that it fired, the Italian Carcano rifle that debuted in that year was the first in a series of bolt-action weapons designed by engineer Salvatore Carcano at the Turin Army Arsenal near the end of the nineteenth century. A short rifle and various carbines were also produced between 1890 and 1945, and approximately three million were eventually completed.

The Mannlicher-Carcano utilized a six-round en bloc clip loaded into an integral magazine, and its effective range was up to 600m (656yds). The rifle was widely used by the Italian Army during World War I and by some German troops as well, while it was fielded by Finnish troops during the Winter War of 1939–40 and by Italian troops in the Mediterranean Theatre during World War II. The Japanese Imperial Navy ordered the Mannlicher-Carcano to fill a need when all domestic Arisaka production was diverted to the Japanese Army after the invasion of China in the 1930s. Variants were chambered for the Carcano 7.35mm (.28in), Mauser 7.92mm (.312in), and Arisaka 6.5mm (.25in) cartridge in addition to the original 6.5mm Mannlicher-Carcano.

In 1888, the bolt action Lee-Metford rifle was designated to replace the Martini-Henry in the ranks of the British Army, combining an eight- or 10-round magazine and rear locking bolt system devised by James Lee with William Metford's rifled barrel with seven interior grooves. The Lee-

Metford endured nine years of research and development before it was deemed acceptable by the British military establishment.

The official tenure of the Lee-Metford was relatively short, however, and it was destined for replacement by the Lee-Enfield in 1895, but the replacement process itself was lengthy, and the Lee-Metford remained in service through the First and Second Boer Wars and into World War I. The 7.7mm (.303in) Lee-Metford fired the .303 Mark I cartridge and was capable of a rate of fire of up to 20 rounds a minute, effective to 730m (798yds). A few Lee-Metford rifles were converted to semiautomatic weapons on a trial basis.

The Winchester Repeating Arms Company was one of the offspring of the earlier Volcanic Repeating Arms Company begun with the partnership of Horace Smith and Daniel Wesson, later of Smith & Wesson fame. After

LEE-METFORD
COUNTRY OF ORIGIN
United Kingdom
DATE
1888
CALIBRE
7.7mm (.303in)
WEIGHT
4.2kg (9.4lb)
OVERALL LENGTH
1257mm (49.5in)
FEED/MAGAZINE
Bolt action repeater
RANGE
730m (798yds)

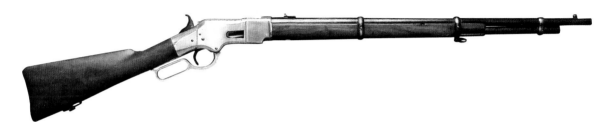

the departure of the founders, Volcanic became insolvent and was purchased
by Oliver Winchester. His talented engineer, Benjamin Tyler Henry,
subsequently perfected the legendary Henry rifle, and Winchester named the
company after himself.

Gun that Won the West

The first repeating rifle bearing the Winchester name was the Model 1866,
which gained everlasting fame during the settlement of the western United
States. The Model 1866 is known to history as the 'Gun that Won the West.'
Originally chambered for the 11.1mm- (.44in-) calibre Henry cartridge,
the Model 1866 corrected several problems with the Henry rifle such as an
overheating barrel and difficulty reloading. It was nicknamed 'Yellow Boy'
due to the bronze sheen of its gunmetal receiver. The lever action rifle was
fed by a 15-round tube magazine, and about 170,000 were manufactured
between 1866 and 1898.

For many, the Winchester Model 1873 shares the moniker of the 'Gun
that Won the West' along with the Model 1866. This 11.1mm- (.44in)
calibre rifle was later also chambered for the 9.6mm- (.38in-) calibre
cartridge and allowed an individual to conveniently carry the same type
of ammunition for both rifle and handgun. The Model 1873 included a
stronger frame and a dust cover over its moving parts. Colt soon began
manufacturing its Army issue revolver in matching 11.1mm (.44in) calibre,
and the Colt Frontier Model became immensely popular. The Winchester
Model 1873 was also produced as a musket and carbine, and by 1919 over
720,000 had been manufactured.

Before the turn of the twentieth century, Winchester produced several
improved versions of its iconic repeating rifle, including the Model 1876
that was chambered to accommodate the heavier cartridges beginning to
appear and incorporated a stronger and larger receiver. President Theodore
Roosevelt was particularly fond of his 1876 'Centennial' Model, nicknamed
for the 100th anniversary of the United States.

WINCHESTER 1866

COUNTRY OF ORIGIN
United States
DATE
1866
CALIBRE
11.1mm (.44in)
WEIGHT
4.3kg (9.5lb)
OVERALL LENGTH
1252mm (49.3in)
FEED/MAGAZINE
Lever action repeater
RANGE
91m (100yds)

LEFT: **An avid hunter, a
youthful future President
Theodore Roosevelt is
pictured wearing buckskin
and carrying one of his
favourite Winchester
repeating rifles.**

WINCHESTER 1894
COUNTRY OF ORIGIN
United States
DATE
1894
CALIBRE
8.13mm (.32in)
WEIGHT
3.1kg (6.8lb)
OVERALL LENGTH
960mm (37.8in)
FEED/MAGAZINE
Lever action repeater
RANGE
91m (100yds)

TARPLEY CARBINE
COUNTRY OF ORIGIN
Confederate States of America
DATE
1863
CALIBRE
13.2mm (.52in)
WEIGHT
3.4kg (7.5lbs)
OVERALL LENGTH
1016mm (40in)
FEED/MAGAZINE
Single shot
RANGE
91m (100yds)

The Winchester Model 1886 accommodated still heavier ammunition and was fitted with locking block action rather than the toggle configuration of the Model 1876. The Model 1892 reverted to the original Winchester premise of firing shorter rounds that could also be used by the handguns of the day. This rifle was manufactured into the 1930s. The Winchester Model 1894 was the company's first rifle intended for use with smokeless powder. Designed by the legendary John Browning, sales of the Model 1894 exceeded seven million during a production history of well more than a century that ended in 2006. Originally chambered for a 8.13mm (.32in) calibre cartridge, it has been produced for several others as well. Other notable Winchesters, some in differing calibres, were the Model 1885, 1895, 1903, 1905 and the modern Models 88 and 9422.

Numerous other firearms manufacturers produced repeating rifles that were associated with America's westward expansion. Among the best known were the Colt Lightning, Marlin Model 93, Remington Keene repeating rifle and the Savage Model 1899.

Cartridge Carbines
The diversity of long arms during the nineteenth century produced a degree of specialization that is most familiar with the introduction of the carbine, a long arm that is generally smaller in size and with a shorter barrel than the standard infantry rifle or musket. The carbine was, therefore, lighter and easier to handle and found favour with cavalry, artillery and other non-infantry units, including officers and staff personnel.

One of the few rifles that originated within the Confederacy during its brief existence was the Tarpley Carbine, developed by Jere H. Tarpley and manufactured in the state of North Carolina from 1863 to 1864. The breechloading Tarpley fired a 13.2mm- (.52in-) calibre cartridge and was similar in that respect to the Sharps carbine. Only about 400 were completed during the American Civil War.

The Berthier Artillery Musketoon Models 1890 and 1892 were carbines based on the Berthier rifle, destined for general issue to the French Army. The bolt action Berthier carbine was an improvement on earlier single shot carbines that had proven difficult to load in action and particularly on horseback. It fired the same 8mm (.314in) cartridge as the Berthier rifle.

The light, sturdy Spencer carbine proved its worth during the American Civil War. Particularly popular with Union cavalry, these weapons enabled troopers to often hold larger formations of enemy soldiers at bay with sheer firepower. The Spencer carbine, like the larger Spencer repeating rifle, fired the 13.2mm- (.52in-) calibre Spencer rimfire cartridge from a seven-round magazine situated in the buttstock. During the Civil War, the U.S. government purchased more than 95,000 Spencer carbines.

Following the defeat of France in the Franco–Prussian War, the Prussian government came into possession of large numbers of Chassepot rifles, the fine frontline long arm of the French Army for more than a decade. The Prussians, along with the armed forces of other German states, cut down many of these captured rifles to shorter carbine size and rechambered them for the 11mm (.43in) Mauser cartridge. These were in service with German artillery and cavalry units during the 1880s.

With its 450mm (17.7in) barrel, the Moschetto 1891 per cavalleria, or Model 1891 Cavalry carbine, was based on the Italian Carcano rifle in service into the early twentieth century with the Italian Army. With its barrel nearly 102mm (4in) shorter than the standard rifle, the Moschetto was equipped with an integral bayonet and fired the same 6.5mm (0.25in) cartridge.

SPENCER CARBINE
COUNTRY OF ORIGIN
United States
DATE
1860
CALIBRE
13.2mm (.52in)
WEIGHT
3.63kg (8lb)
OVERALL LENGTH
1067mm (42in)
FEED/MAGAZINE
Lever action repeater
RANGE
457m (500yds)

World War I

On 28 June 1914, Bosnian Serb and nationalist Gavrilo Princip fired several shots with a Browning FN Model 1910 blowback-operated semiautomatic pistol. His targets were Archduke Franz Ferdinand of Austria-Hungary and his wife Sophie, Duchess of Hohenberg. In a short time both victims were dead.

LEFT: British soldiers line a trench on the bleak Western Front in World War I, awaiting the order to go over the top and confront the enemy across No Man's Land.

The double assassination in Sarajevo triggered World War I as Austria-Hungary declared war on Serbia, and the nations of Europe, bound by treaties and alliances, were obliged to enter the conflict. Although Princip had ignited the Balkan tinderbox with a pistol and the world would learn the deadly killing power of the machine gun, massed artillery, the airplane, the submarine and poison gas, the principal weapon of the foot soldier from 1914 to 1918 was the bolt-action, magazine-fed rifle.

Improvements in the design and durability of the rifle and such innovations in ammunition as the aerodynamic jacketed bullet brought the shoulder arm to a new level of deadly efficiency in the hands of a well-trained and motivated infantryman. Although many soldiers of the Great

LEBEL M1886
COUNTRY OF ORIGIN
France
DATE
1886
CALIBRE
8mm (.314in)
WEIGHT
4.41kg (9.7lb)
OVERALL LENGTH
1300mm (52in)
FEED
Bolt action repeater
RANGE
400m (438yds)

ABOVE: Archduke Franz Ferdinand of Austria-Hungary and his wife, Princess Sophie of Hohenberg, walk toward their open car in Sarajevo. On 28 June, 1914, the two were assassinated, sparking World War I.

War were conscripts, depending upon the specific weapon employed, the skilled rifleman and his comrades were capable of delivering a veritable whirlwind of lethal fire at an advancing enemy.

Western Bolt Action Rifles

The French Army struggled to develop and supply its troops in the field with an adequate, modern rifle throughout World War I. The standard issue French rifle of 1914–15 was the 8mm (.314in) Lebel design of 1886. The Lebel entered service in the spring of 1887, and in short order it became apparent that the rifle's magazine functioned in a less than optimal manner. The magazine held eight rounds, at least comparable to other rifles of its type; however, the time-consuming loading process required cartridges to be fed end to end through a tube placed at the forward end of the weapon. Obviously, therefore, a French soldier attempting to reload his Lebel M1886 in combat was quite vulnerable. Further compromising its performance was a shifting centre of gravity as successive rounds were fired, resulting in a lack of sustained accuracy.

The first rifle to use smokeless powder, the Lebel M1886 was essentially rendered obsolete by 1915 and was designated for replacement by the lighter Lebel Berthier Models 1902 and 1907, which were never produced in great

quantity but were fed by a three-round clip that eliminated the troublesome tube magazine. Due to the shortage of more modern rifles, the M1886 remained the primary infantry shoulder arm for French forces throughout World War I.

French Foreign Legion

More than 400,000 examples of a modified M1907, the Fusil Berthier Mle 1907/15, were manufactured and issued to French colonial troops and soldiers of the French Foreign Legion. The standard Lebel Berthier models, however, remained limited in number, while the three-round magazine was proven inadequate in combat. In 1916, another modified Berthier, the Fusil Mle 1907/15-M16, with a five-round magazine, was authorized. This improved model did not reach frontline units of the French Army until the summer of 1918, much too late to affect the progress of the war on the Western Front. A number of French officers, meanwhile, insisted on retaining the aging Lebel rifle, perhaps due to its eight-round capacity.

After the Berthier models entered service, the Lebel continued to equip frontline French Army units as a primary weapon and as a sniper rifle mounting high-powered sights.

Another French attempt to replace the Lebel M1886 resulted in the development of the Fusil Automatique Modele 1917, which actually became operational with the French Army in the spring of 1916. The rifle was also known as the Modele 1917 RSC in reference to the design collaboration of Ribeyrolles, Sutter and Chauchat that produced it. By the time the war

LEBEL BERTHIER 1907/15
COUNTRY OF ORIGIN
France
DATE
1915
CALIBRE
8mm (.314in)
WEIGHT
3.8kg (8.4lb)
OVERALL LENGTH
1306mm (51.4in)
FEED/MAGAZINE
Bolt action; 3-round clip
RANGE
2000m (2187yds)

FUSIL 1917
COUNTRY OF ORIGIN
France
DATE
1916
CALIBRE
8mm (.314in)
WEIGHT
5.28kg (11.64lb)
OVERALL LENGTH
1330mm (52.4in)
FEED/MAGAZINE
Bolt action, 5-round clip;
internal box
RANGE
1200mm (1312yds)

ABOVE: **President Woodrow Wilson and General John J. Pershing review French soldiers turned out in full uniform. These troops stand at attention with bayonets fixed to their Lebel rifles.**

ended in November 1918, more than 80,000 examples of the Modele 1917 had been manufactured at the French government's Manufacture d'armes de Saint-Étienne facility.

The bolt action Modele 1917 utilized an internal box magazine that was fed by a five-round clip. It fired an 8mm (.314in) cartridge, and a long recoil and gas-operated ejection for spent cartridges did provide some rudimentary aspects of semiautomatic operation. Still, those French troops who received the Modele 1917 were less than impressed with its performance, even though it appeared late in the war and a relative few were actually issued. One significant drawback was that its extended length of 1330mm (52.4in) was awkward for steady aiming and for close quarter fighting that was sometimes experienced during trench warfare.

In 1918 the rifle's proprietary ammunition clip was changed to the Berthier clip, in universal service with other French rifles. Only about 4000 of the 1918 variant were produced, and these were fielded into the 1920s.

Two Berthier carbines, the Model 1907 and the Mousqueton Berthier Mle 1892/M16, performed well during World War I. The Model 1907 replaced the old Mle 1874 Gras single-shot carbine, and some soldiers preferred it to the Lebel M1886 since it was manufactured with a single-piece stock and utilized a three- or five-round clip rather than the tubular magazine. With a five-round charger-loaded magazine, the Model 1892/16 was highly successful and deployed with light infantry, cavalry and specialized units.

Designed by the Mauser brothers and manufactured by the arsenals of Imperial Germany and numerous private contractors, the Mauser Gewehr 1898 rifle replaced the Gewehr 1888, a small-bore rifle that used smokeless powder and was intended to keep pace with the advances of the French Lebel M1886, and it became the standard German shoulder arm of World War I. During a 20-year period from 1898 to 1918, approximately five million of the Gewehr 98 were manufactured, and it remained in service with the Volkssturm home defence forces during World War II after it was withdrawn from service with regular German Army troops in 1935 in favour of the Karabiner 98k.

The Mauser Gewehr 98 fired a 7.92mm (.312in) cartridge from a five-round internal box magazine that was fed by a brass stripper clip. The weapon was relatively easy to load since the stock was cut down on the right-hand side to allow the rifleman to insert the cartridge more rapidly and safely than with the downward motion of the thumb that was necessary with the

MOUSQUETON 1892
COUNTRY OF ORIGIN
France
DATE
1892
CALIBRE
8mm (.314in)
WEIGHT
3.1kg (6.8lb)
OVERALL LENGTH
945mm (37.2in)
FEED
Bolt action; 3-round charger clip
RANGE
2000m (2187yds)

MAUSER GEWEHR 1888
COUNTRY OF ORIGIN
Germany
DATE:
1898
CALIBRE
7.92mm (.312in)
WEIGHT
4.09kg (9lb)
OVERALL LENGTH
1250mm (49.2in)
FEED
Bolt action; 5-round clip; internal magazine
RANGE
550m (601.49yds)

MAUSER GEWEHR 1898

COUNTRY OF ORIGIN
Germany
DATE
1898
CALIBRE
7.92mm (.312in)
WEIGHT
4kg (9lb)
OVERALL LENGTH
1250mm (49.2in)
FEED/MAGAZINE
Bolt action; 5-round clip;
internal magazine
RANGE
550m (601.49yds)

contemporary British Lee-Enfield rifle. Although it was prone to jamming in the presence of excessive dust and debris, or prolonged combat use when opportunities to clean and service it were limited, the bolt action of the Gewehr 98 worked well in the field.

For the infantryman, operating the heavy bolt required a strong hand in combat. The soldier was often obliged to realign his vision through the sights after operating the mechanism. Well-suited for the mobile warfare experienced during the opening months of World War I, the Mauser Gewehr 98 became somewhat problematic with the advent of warfare in the trenches of the Western Front. The weapon was rather ponderous and somewhat heavy with an unloaded weight of 4kg (9lb) and a length of 1250mm (49.2in). An experienced German infantryman was capable of firing up to 12 rounds a minute despite the fact that the five-round magazine of the Gewehr 98 was of lower capacity than the British Lee-Enfield.

During World War I the sniper came into his own, both with the Allied armies and those of the Central Powers. In the spring of 1915, more than 15,000 of the Mauser Gewehr 98 were ordered to be equipped with telescopic sights to employ them as highly accurate sniper rifles. When the war ended in 1918, that number had been significantly exceeded with the addition of the telescopic sights completed on more than 18,000 of the sniper variant. Prior to World War I, a shortened version of the Gewehr 98, the Karabiner 98a, was produced as a cavalry weapon. However, this experiment was disappointing and production discontinued.

Germany Enters the War

In August 1914, just days after Germany entered World War I on the side of Austria-Hungary, a young lieutenant of the 6th Wurttemberger Regiment was in action at the Belgian village of Bleid. Erwin Rommel was destined to command the vaunted Afrika Korps in the Desert War, rise to the rank of field marshal and earn the nickname of the Desert Fox during World War II. As a junior officer on this day, he carried a Mauser Gewehr 98 into combat.

'I quickly informed my men of my intention to open fire. We quietly released the safety catches; jumped out from behind the building; and standing erect, opened fire on the enemy nearby. Some were killed or wounded on the spot; but the majority took cover behind steps, garden walls and woodpiles and returned our fire. Thus, at very close range, a very hot firefight developed. I stood taking aim alongside a pile of wood. My adversary was twenty yards ahead of me, well covered, behind the steps of a house. Only part of his head was showing. We both aimed and fired almost at the same time and missed. His shot just missed my ear. I had to load fast, aim calmly and quickly, and hold my aim. That was not easy at twenty yards with the sights set for 400 meters, especially since we had not practiced this type of fighting in peacetime. My rifle cracked; the enemy's head fell forward on the step.'

During World War I, the Belgian solder was typically equipped with a slightly modified version of the German Mauser rifle designated the Fusil FN-Mauser Modele 1889. This weapon had been manufactured in Belgium by the Fabrique Nationale d'Armes de Guerre, a company formed in 1889 to manufacture the Mauser rifle for the Belgian government. The development of the Model 1889 had begun in the early 1880s, and the weapon was similar to the 1871 series, the first rifles produced by Mauser.

LEFT: A German infantryman, possibly a sniper moving into position, lies prone on the battlefield. With added optics, the Mauser Gewehr 98 was a highly accurate sniper rifle.

FUSIL MLE 1889
COUNTRY OF ORIGIN
Belgium
DATE
1889
CALIBRE
7.65mm (.301in)
WEIGHT
4kg (9lb)
OVERALL LENGTH
1250mm (49in)
FEED/MAGAZINE
Bolt action; 5-round vertical
magazine
RANGE
550m (601.49yds)

The bolt-action FN-Mauser Modele 1889 was distinguished by its five-round vertical magazine, barrel shroud and heavy wooden stock. It fired a 7.65mm (.301in) cartridge, and its box magazine was a considerable improvement over the tubular magazines of older weapons. With the rifle's open action, the magazine was loaded using a charger.

When the original Mauser Model 1889 failed to impress the German military establishment in field tests against a rifle produced by Mannlicher of Austria, the hopes of a government contract for the Model 1889 evaporated. However, the Belgian attaché had had been impressed with the

rifle's performance during the Bavarian Arms Trials of 1884 and became an advocate for its adoption by the Belgian military.

Subsequently, the Fabrique Nationale (FN) was formed to produce the Model 1889 in Belgium. When Belgian production facilities failed to produce the rifle in adequate numbers, the Belgians negotiated a contract with a British company to produce the Model 1889. By the time production ceased, those made in Britain and Belgium numbered 250,000 or more.

The Turkish Army carried the Mauser Gewehr 98 and the modified Mauser Model 1889 during World War I. The FN Mauser Model 1889 was delivered to the Turks in completion of an order that had been originally filled with the earlier Mauser Model 1887. A clause in the Turkish contract with Mauser stipulated that if another nation purchased an improved version of the Model 1887 the Turks would receive the upgraded weapon to complete their deliveries. When Belgium purchased the Model 1889 and began its manufacture, the clause in favour of the Turks was triggered.

ABOVE: The camp of Turkish troops at Gallipoli reveals the harsh terrain encountered during the bloody 1915 campaign. Turkish soldiers carried Mauser rifles, and several are seen stacked in the foreground.

MAGAZINE LEE ENFIELD
COUNTRY OF ORIGIN
United Kingdom
DATE
1895
CALIBRE
7.7mm (.303in)
WEIGHT
4kg (8.8lb)
OVERALL LENGTH
1130mm (44.49in)
FEED/MAGAZINE
Bolt action; 10-round detachable box magazine
RANGE
500m (550yds)

SMLE MK I
COUNTRY OF ORIGIN
United Kingdom
DATE
1904
CALIBRE
7.7mm (.303in)
WEIGHT
4.19kg (9.24lb)
OVERALL LENGTH
1130mm (44.49in)
FEED/MAGAZINE
Bolt action; 10-round detachable box magazine
RANGE
500m (550yds)

Making its debut in the autumn of 1895 as the 7.7mm- (.303in-) calibre, Rifle, Magazine, Lee-Enfield, the original version of the iconic Lee-Enfield rifle and its offspring equipped the soldiers of the British and Commonwealth armies for more than half a century before their replacement as standard issue in 1957. The longevity of the Lee-Enfield rifles lay in their excellent design, truly one of the best basically configured bolt action rifles in history.

The Lee-Enfield was developed substantially from the combination of the best attributes of the earlier Lee-Metford rifle, particularly its outstanding bolt action that was a product of collaboration between designers James Paris Lee and William Metford, with an excellent rifling that was introduced by the Royal Small Arms Factory located in Enfield Lock, a borough of North London.

Smokeless Cordite

The Lee-Enfield was also developed for use with new smokeless cordite powder that eliminated the great plume of smoke produced by earlier propellants, and better concealed the soldier on the battlefield after firing his rifle while also reducing the buildup of residue that fouled earlier rifle barrels and resulted in more frequent maintenance.

The initial production version of the Magazine Lee-Enfield was in service from 1895 to 1907, featuring the new rifling and improved sights. It was among those firearms utilized by British troops during the Second Boer War in South Africa.

In January 1904, the Short Magazine Lee-Enfield Mk I was issued and quickly became referred to as the 'SMLE', which was destined to become the standard British rifle design for the next 50 years. The hallmark of the Mk I SMLE was its blunt nose, with only a fraction of the barrel extending beyond the stock and the boss for the bayonet protruding beyond the muzzle as approximately 127mm (5in) had been removed from the length of the earlier Lee-Enfield barrel. The Short Magazine Lee-Enfield Mk II, in service from 1906 to 1927, featured only slight differences from the Mk I, including enhanced sights and barrel. It could also be reloaded with a charger clip. The fact that the barrels of the SMLE Mk I and II were shorter than the original Magazine Lee-Enfield created some initial controversy due to concerns that accuracy would be adversely affected due to an exaggerated recoil.

Introduced in 1907, the Lee-Enfield Rifle No.1 Mk III SMLE earned a reputation as one of the truly outstanding infantry rifles of the twentieth century. The famous 7.7mm- (.303in-) calibre Mk III incorporated a 10-round box magazine that could be loaded from the top with five-round chargers. The bolt action of the Mk III was so smooth that a rifleman could maintain his focused sight picture and acquire targets while operating the rifle. The more stable shooting perspective resulted in greater accuracy. In contrast, the German soldier worked a rough bolt action with the Mauser Gewehr 98 and was required to reorient himself to the task at hand following each shot fired.

SMLE MK II
COUNTRY OF ORIGIN
United Kingdom
DATE
1906
CALIBRE
7.7mm (.303in)
WEIGHT
4.19kg (9.24lb)
OVERALL LENGTH
1130mm (44.49in)
FEED/MAGAZINE
Bolt action; 10-round detachable box magazine
RANGE
500m (550yds)

SMLE MK III
COUNTRY OF ORIGIN
United Kingdom
DATE
1907
CALIBRE
7.7mm (.303in)
WEIGHT
3.96kg (8.73lb)
OVERALL LENGTH
1132mm (44.57in)
FEED/MAGAZINE
Bolt action; 10-round detachable box magazine
RANGE
500m (550yds)

British infantrymen shoulder their SMLE rifles. The SMLE No. 1 MK III became an iconic rifle of the 20th century.

Sustained Firepower

Concentrated and massed infantry rifle fire was a hallmark of British Army tactics; therefore, the Mk III SMLE was considered an integral component of overall British battlefield strategy. The rifle proved to be ideal for such operations. A well-trained and motivated British soldier was typically capable of achieving a sustained rate of fire of 15 rounds a minute.

During the thickest of the fighting, British soldiers regularly referenced the 'mad minute', a period of sustained fire during which an individual soldier could load, aim and fire up to an astounding 30 rounds a minute. Actually, the German high command received some reports from Western Front locations, where the fighting was most intense, that its infantry had been under fire from an incredibly high number of British machine guns. The enemy troops actually confronting those German soldiers were, in fact, regular British infantrymen firing their Mk III rifles.

The Lee-Enfield Mk III SMLE had a sighted range of around 100m (109yds), and it proved durable on the battlefield, withstanding the rigours of heavy use and the wet, muddy confines of the trenches of the Western Front. The quality of the rifle was so outstanding that approximately three million were produced, and some still remain in service today with armies around the globe.

The basic Mk III rifle was modified several times through the years, and a sniper variant served with Commonwealth forces into the 1990s. The Lee-Enfield No.4 Mk 1 became the standard issue British infantry rifle of World War II, while the Lee-Enfield No.5 Mk I, a shortened version of the standard No.4 rifle, was issued in 1943. Commonly referred to as the Jungle Carbine, the performance of the No.5 Mk I was disappointing.

SMLE MK III (WITH GRENADE LAUNCHER)
COUNTRY OF ORIGIN
United Kingdom
DATE
1916
CALIBRE
7.7mm (.303in)
WEIGHT
3.93kg (8.625lb)
OVERALL LENGTH
1133mm (44.6in)
FEED/MAGAZINE
N/A
RANGE (GRENADE)
100m (109yds)

1914 ENFIELD
COUNTRY OF ORIGIN
United Kingdom/United States
DATE
1914
CALIBRE
7.7mm (.303in)
WEIGHT
4.35kg (9.6lb)
OVERALL LENGTH
1175mm (46.2in)
FEED/MAGAZINE
Bolt action: 5-round box magazine
RANGE
500m (550yds)

1917 ENFIELD
COUNTRY OF ORIGIN
United Kingdom/United States
DATE
1917
CALIBRE
7.62mm (.30in)
WEIGHT
4.17kg (9.2lb)
OVERALL LENGTH
1175mm (46.25in)
FEED/MAGAZINE
Bolt action; 6-round magazine,
5-round clip
RANGE
500m (550yds)

ROSS RIFLE
COUNTRY OF ORIGIN
Canada
DATE
1903
CALIBRE
7.7mm (.303in)
WEIGHT
4.48kg (9.875lb)
OVERALL LENGTH
1285mm (50.6in)
FEED/MAGAZINE
Bolt action; 5-round magazine
RANGE
500m (550yds)

An estimated 17 million examples of all variants of the Magazine Lee-Enfield and the Short Magazine Lee-Enfield have been manufactured.

Although it was originally configured to fire a high-powered round, a rifle known as the Pattern 1914 Enfield was eventually adapted to the 7.7mm-(.303-) calibre ammunition generally in use. It was manufactured under contract by companies in the United States and never reached the front in great numbers. Fashioned from the Pattern 1913 Enfield, which was itself based on the Mauser Gewehr 98, the Pattern 1914 is not a part of the Lee-Enfield series and should not be confused with those weapons.

The Enfield Model 1917, sometimes referred to as the American Enfield, was adapted from the Pattern 1914 Enfield and manufactured in the United States from 1917 to 1918. More than 2.1 million were manufactured, and the Model 1917 continues to serve today as a ceremonial rifle. Both Remington and Winchester produced the Pattern 1914 and Pattern 1917, which was eventually reconfigured to accept the American .30-06 rifle cartridge.

Political Disagreement

A minor political disagreement broke out between the governments of Great Britain and Canada around the turn of the twentieth century when the British refused to license the Magazine Lee-Enfield rifle for production in Canada. Sir Charles Ross, a well-connected patrician, had developed a straight-pull rifle and presented his alternative to the British Lee-Enfield to the Canadian government. The Ross Rifle performed well during trials and on the rifle range, and the first contract for the delivery of 12,000 of them was signed in 1903.

The Ross was chambered for the 7.7mm- (.303in-) calibre round and fed by a five-round magazine. Its straight pull action differed from the more common bolt action of the contemporary Mauser and Lee-Enfield designs. In the Ross, the locking lugs were mounted on a screw that rotated into place in the rifle's receiver. The Ross was theoretically capable of generating a rate of fire up to 20 rounds a minute given that its operator was not required to rotate the handle a quarter turn to disengage the bolt.

Mechanical issue plagued the Ross following the delivery of the first 1000 rifles to a unit of the Royal Northwest Mounted Police for field-testing.

BELOW: Canadian soldiers clean and repair their Ross rifles somewhere in France. The Ross proved to be a great disappointment in the field due to mechanical issues.

The rifle jammed regularly and was susceptible to fouling from the normal wear and tear of field use and the grit and grime that accumulated. In all, more than 100 deficiencies were found in the rifle, but most of these were thought to have been corrected before it was issued to troops in the field during World War I. However, its performance remained disappointing. In addition to the mechanical troubles encountered with the Ross Rifle, its length was problematic in the confines of the trenches.

By the summer of 1916, a controversy raged over the retention of the Ross rifle in the ranks of the three Canadian infantry divisions deployed in Europe. Subsequently, the decision was made to replace it with the Lee-Enfield. Despite its shortcomings, the Ross was highly accurate and was retained in service with Canadian snipers for the duration of the war. Approximately 500 Ross rifles were equipped with telescopic sights for sniper use and designated the Mk III.

Eventually about 420,000 Ross rifles were manufactured, and more than 300,000 of these were purchased by Great Britain. A sporting variant, the 1897 Magazine Sporting Rifle, was produced in the United States after Ross established a manufacturing facility in Hartford, Connecticut.

At the turn of the twentieth century, the military establishments of a number of nations were searching for a durable and accurate standard issue rifle. Among them was the United States, which held a competition on New York's Governors Island in 1892. Rifle manufacturers and designers from numerous countries entered, and a total of 53 models from such notables as Mauser, Lee, Mannlicher and the top American firms were put to the test.

Krag-Jorgensen

None of the three finalists were American companies, and the contract was awarded to Krag, a Norwegian designer that bested the two other finalists, Lee and Mauser. The Krag-Jorgensen bolt-action repeating rifle traces its origins to the 1860s and was the culmination of the collaborative efforts of Ole Krag, a captain in the Norwegian Army, and gunsmith Erik

KRAG-JORGENSEN
COUNTRY OF ORIGIN
Norway
DATE
1886
CALIBRE
7.62mm (.30in)
WEIGHT
3.375kg (7.4lb)
OVERALL LENGTH
986mm (38.8in)
FEED/MAGAZINE
Bolt action; 5-round magazine
RANGE
500m (550yds)

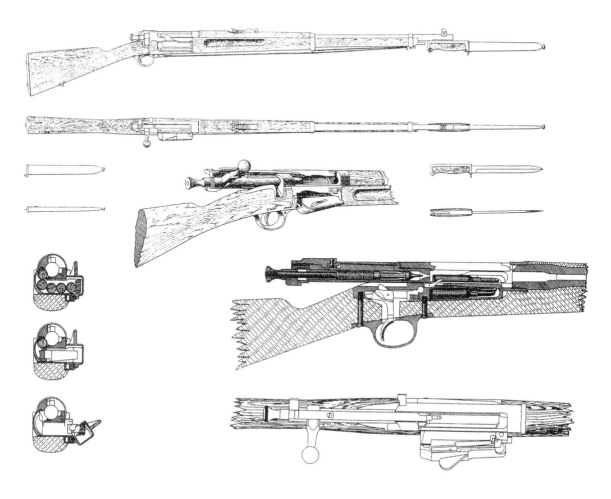

Jorgensen. The rifle was adopted by the Danish Army in 1889 and, following competitive trials, by the Norwegian Army in 1894. The Krag-Jorgensen entered service with the U.S. Army in 1894 as well and served as its frontline shoulder arm for the next nine years.

Unlike other bolt-action rifles of the period, the Krag-Jorgensen was loaded with individual cartridges through an opening in the side of the integral receiver. Other rifles utilized a stripper or charger clip to load via a box magazine. In the Krag-Jorgensen it was theorized that the rifle could be loaded in combat quickly without opening the bolt, a task that would otherwise momentarily take the rifleman out of action. The slower single round loading operation was also thought to conserve ammunition. The Danish Krag-Jorgensen differed from the U.S. version in that it was loaded by a hinged magazine door that opened forward.

ABOVE: **This detail of the Krag-Jorgensen reveals the various components of the Norwegian designed rifle. The U.S. Army adopted a modified version of the Krag-Jorgensen in 1894.**

The Norwegians adopted the rifle to fire their recently designed 6.5mm (.25in) rimless cartridge. The Danish version fired a 7.87mm (.31in) rimmed cartridge, and the American round was the .30-40 Krag, the first smokeless cartridge placed in service by the U.S. Army. The rifle's effective range was 900m (984yds), and its magazine originally had a capacity of 10 rounds, later modified to five rounds. A magazine cutoff allowed the rifleman to load and fire single rounds while the magazine remained fully loaded. In the event that it was necessary, a switch could be flipped to engage the magazine to engage a rapidly growing battlefield threat.

In the hands of American troops, the Krag-Jorgensen was deployed during the Spanish-American War and the Philippine Insurrection. Its slower loading operation proved detrimental in combat against the German-manufactured Mauser rifles then in use with Spanish infantry.

Numerous variants to the original Krag-Jorgensen were produced, including carbines and sniper rifles. Its service life among various armed forces lasted nearly 60 years – until the end of World War II. Production of all types exceeded 700,000. The smooth bolt action of the Krag-Jorgensen is favoured by marksmen today, and the rifle is a collectors' favourite.

When the Krag-Jorgensen proved deficient as the primary combat rifle of the U.S. Army, the American military was already familiar with the merits of the Mauser Gewehr 98. In fact, the army had purchased some quantities of these rifles, and its infantry units had faced the Mauser in action against Spanish troops in Cuba. By the time the American Expeditionary Force arrived in France in 1917, its primary shoulder arm was the Springfield Model 1903. Manufactured at the Springfield Arsenal in Massachusetts, the Model 1903 was a close facsimile in many ways to the battlefield proven Mauser.

Although there were a few minor alterations to the patented Mauser design, the basic weapon varied little from its German configuration. In fact, Mauser eventually filed suit against the U.S. government and received royalty payments following a $3 million settlement for patent infringement.

Springfield Model 1903

The Springfield 1903 entered service in 1905 and was not officially replaced until the M1 Garand was adopted by the U.S. Army in 1937. It was smaller than other rifles of the day but was highly accurate and durable, remaining popular with American soldiers for more than three decades. Its sniper version is in use to this day, and the M1903A4, which made its battlefield

debit in 1942, was the first attempt by the U.S. Army to perfect and
standardize a true sniper rifle.

The Springfield Model 1903 fired the 7.62mm (.30-03 then .30-06)
cartridge, and a veteran combat soldier was capable of firing an average of
15 rounds a minute with ammunition fed from a five-round stripper clip to an
internal box magazine. The rifle was originally in service to fire the flat-ended
7.62mm-calibre M1903 cartridge. However, practical experience during World
War I fueled an effort to develop a heavier bullet. The result was the pointed
M1906 7.62mm-calibre round, with which the rifle has become famous.

Production at the Springfield and Rock Island, Illinois, arsenals topped
800,000 prior to World War I. Eventually several million were manufactured
until production ceased in 1965. Despite some difficulties with its basic
sights, the Springfield Model 1903 was popular for so long simply due
to its accuracy, its durability in difficult conditions, its ease of carry and
its adaptability to carbine configuration and sniper usage. The rifle also
continues to be immensely popular with firearms collectors, marksmen and
hunters, and is often seen in military drill settings.

During World War I, infantry units of the Italian Army were generally
equipped with the Mannlicher Carcano Model 1891 bolt-action rifle,
which, not surprisingly, included features of the German Mauser combined
with design elements of the Austrian Mannlicher firm, particularly the
ammunition clip and feed system. Also known as the Fucile Modelo 1891,
the rifle, developed primarily by Italian engineer Salvatore Carcano, was
initially chambered for a 6.5mm (.256in) rimless cartridge that was fed by
en bloc charger clips into a six-round box magazine. The Model 91 operated

SPRINGFIELD 1903
COUNTRY OF ORIGIN
United States
DATE
1903
CALIBRE
7.62mm (.30in)
WEIGHT
3.9kg (8.625lb)
OVERALL LENGTH
1115mm (43.9in)
FEED/MAGAZINE
Bolt action; 5-round stripper
clip, box magazine
RANGE
750m (820.2yds)

MANNLICHER CARCANO 1891
COUNTRY OF ORIGIN
Italy
DATE
1891
CALIBRE
6.5mm (.256in)
WEIGHT
3.8kg (8.375lb)
OVERALL LENGTH
1291mm (50.79in)
FEED/MAGAZINE
Bolt action; 6-round box
magazine
RANGE
1000m (1093.6yds)

MOSCHETTO 1891
COUNTRY OF ORIGIN
Italy
DATE
1891
CALIBRE
6.5mm (.256in)
WEIGHT
3kg (6.6lb)
OVERALL LENGTH
920mm (36.2in)
FEED/MAGAZINE
Bolt action; 6-round box
magazine
RANGE
600m (656.17yds)

BERDAN RIFLE
COUNTRY OF ORIGIN
United States/Russia
DATE
1869
CALIBRE
10.75mm (.42in)
WEIGHT
4.2kg (9.25lb)
OVERALL LENGTH
1300mm (51.18in)
FEED/MAGAZINE
Berdan I: trapdoor; Berdan
II: bolt action/single shot
breechloader
RANGE
280m (306.21yds)

with a rotating bolt action and produced an adequate rate of fire with an easy loading sequence that was facilitated by a clip that was open and could be loaded into the rifle from either end. The original rifle was standard issue with the Italian Army until 1938 when it was rechambered to fire the heavier 7.35mm (.28in) cartridge.

The Model 1891 was produced in both standard and carbine versions at the manufacturing facility in Turin, Italy, and officially approved for service by the Italian Army in the same year. The carbine was issued primarily to cavalry and specialized units such as alpine troops, and the Moschetto 1891 per Cavalleria with its folding bayonet permanently fixed to the muzzle mounting was produced by several manufacturers, including FNA Brescia.

During the 1930s, the Italians produced about 60,000 examples of the Carcano, under the label of the Type I rifle, for the Japanese Navy. The Turkish armed forces also deployed the Mannlicher Carcano during the period, and it was produced until 1945. Thousands were sold as surplus into the 1960s and 1970s. One of these gained lasting infamy as the weapon used by Lee Harvey Oswald to assassinate U.S. President John F. Kennedy in 1963.

Eastern Bolt Action Rifles

When Austria-Hungary declared war on Serbia, one of the prominent rifles then in use by the Serbian Army was either of the Berdan series developed by American Civil War hero and inventor Hiram Berdan. The series was produced in Russia from the late 1860s through 1891 and included the trapdoor breechblock Berdan I and the single-shot bolt-action Berdan II. Each of these were also used extensively by the Russian Army, surviving in

Italian soldiers negotiate the snow-covered Alps as they move to defensive positions early in World War I.

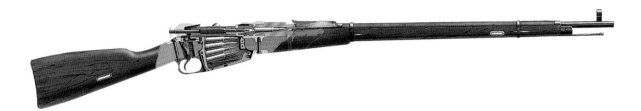

service even after their official replacement by the Mosin Nagant rifle in the early twentieth century.

The armies of Finland and Bulgaria also fielded the Berdan series, and by 1914 Russian manufacturers had exported more than 75,000 to Serbia. In the hands of a highly-trained soldier, the Berdan series rifles were capable of rates of fire of six to eight rounds a minute. The early Berdan rifles were chambered for a heavy 10.75mm (.42in) cartridge.

Mosin-Nagant

Those shoulder arms that were manufactured in Russia were of reasonably good quality but never reached the Czar's frontline troops in sufficient numbers during World War I. One of the most prominent of these was the bolt-action Mosin Nagant Model 1891 designed by Russian Army officer Sergei Mosin and Belgian arms manufacturer Léon Nagant. The Mosin-Nagant Model 1891 was initially manufactured in Belgium and later adopted by the Russian military following a controversial competition between separate designs from Mosin and Nagant.

Although the Mosin-Nagant weapon incorporated some elements suggested by both men, it was never known or recognized in Russia by the combined name that appeared in Western nomenclature. In fact, Nagant eventually filed suit over patent rights to an interrupter that prevented the double feeding and jamming of cartridges in the rifle's receiver – even though he is said to have taken the innovation from Mosin. Since Mosin had won the original competition, he was paid a sum of 200,000 rubles. Mosin, however, could not apply for a patent on the interrupter since he was an officer of the Russian Army.

Weighing the benefits of a continuing association with Nagant, the Russian military establishment decided to award the Belgian 200,000 rubles as well. This decision was made primarily because Nagant was not working with foreign governments at the time and was one of a relative few firearms designers who were eager to share knowledge and expertise with the Russians.

MOSIN-NAGANT 1891
COUNTRY OF ORIGIN
Russia
DATE
1891
CALIBRE
7.62mm (.30in)
WEIGHT
4.37kg (9.625lb)
OVERALL LENGTH
1305mm (51.4in)
FEED/MAGAZINE
Bolt action; 5-round box magazine
RANGE
500m (550yds)

LEFT: Famous for raising and outfitting a regiment of Union sharpshooters during the American Civil War, Colonel Hiram Berdan invented a rifle that became popular in Russia.

STEYR-MANNLICHER 1895

COUNTRY OF ORIGIN
Austria-Hungary
DATE
1895
CALIBRE
8mm (.314in)
WEIGHT
3.8kg (8.36lb)
OVERALL LENGTH
1272mm (50.12in)
FEED/MAGAZINE
Bolt action; 5-round en bloc
clip, box magazine
RANGE
500m (550yds)

The 7.62mm (.30in) bolt action Mosin-Nagant rifle was fed by a five-round box magazine, and the weapon proved to be quite serviceable in the harsh weather conditions often encountered on the Eastern Front. Among later improvements to the weapon, which continued into the 1930s, was the introduction of a feed mechanism that substantially decreased the likelihood of failure during the heat of combat.

The Mosin-Nagant was widely distributed to Russian Army units as manufacturing facilities began production and increased the availability of the rifle. By the beginning of the Russo–Japanese War of 1904–05 nearly four million had been delivered. During World War I, the Russian government ordered an additional 3.2 million rifles from the American firms of Remington and Westinghouse. The Bolshevik Revolution, however, halted these deliveries.

A dragoon rifle and Cossack rifle, both shortened for horsemen, were among the variants of the Mosin-Nagant Model 1891 that entered service with the Russian Army by the 1930s. Later models equipped the Red Army during World War II, and many Mosin-Nagant rifles remained in service for decades. When production ceased in 1965 an estimated 37 million had been made.

Due to the fact that the rifleman was not required to turn the bolt during the reloading process, the Steyr-Mannlicher Model 1895 straight-pull bolt-action rifle proved to be an outstanding battlefield performer during World War I with a higher rate of fire than other contemporary rifles. The Model 1895, developed by Austrian engineer Ferdinand Ritter von Mannlicher, was the standard issue rifle of the Austro-Hungarian Army in the Great War.

In 1903, the Bulgarian government purchased the Steyr-Mannlicher Model 1895, and the rifle equipped most of that nation's infantry units during World War I. Its straight-pull bolt further served as a model for the Canadian M1905 Ross Rifle, which proved to be a major disappointment under the duress of the battlefield.

Although the Model 1895's bolt was a challenge to pull back and offered more resistance to its basic manipulation than other models, the rifle was

capable of firing up to an amazing 35 rounds a minute. It was nicknamed the
'Ruck-Zuck', which translates from the German slang as 'Right Now' or 'Very
Quick.' While the high rate of fire with the straight-pull bolt was a definite
advantage in combat and improved the comprehensive durability of the
weapon, it also resulted in additional required maintenance such as regular
cleaning. Since there was little to assist the bolt action itself in the ejection
of the spent cartridge, and the rifle fired at such a high rate, residue rapidly
built up and had to be cleared.

The Steyr-Mannlicher Model 1895 was loaded via an internal box
magazine by a five-round en-bloc clip. This was later updated to a stripper
clip for use in subsequent variants. The rifle fired a 8mm (.314in) cartridge,
and from 1895 to 1918 more than three million were produced. It remained
in service with the armed forces of numerous countries through the end
of World War II and continues to turn up frequently in the hands of
paramilitary and guerrilla fighters around the world.

Japanese Weaponry

So named due to its entry into service in 1897, the thirtieth year of the Meiji
Dynasty, the Arisaka Type 30 rifle was a bolt-action weapon that served as
the standard Japanese Army infantry shoulder arm through the end of the
Russo–Japanese War in 1905. It replaced the Type 22 rifle, also known as the
Murata, and was considered a significant improvement over its predecessor.
Although it was officially replaced in 1905 by the Type 38 bolt-action rifle,
the Type 30 remained in service throughout World War II.

Chambered for the semi-rimmed Arisaka 6.5mm (.256in) cartridge, the
Type 30 rifle could fire between 10 and 15 rounds a minute from a five-round
internal magazine. Colonel Nariakira Arisaka supervised the manufacture
of the Type 30 and its variants at the Koishikawa Arsenal in Tokyo, and
the series bears his name. Over 550,000 Type 30 rifles were produced during
a relatively short manufacturing run, and approximately 45,000 examples
of a shortened carbine variant that lacked an apparatus to accommodate a

ARISAKA 30
COUNTRY OF ORIGIN
Japan
DATE
1897
CALIBRE
6.5mm (.256in)
WEIGHT
3.95kg (8.7lb)
OVERALL LENGTH
1280mm (50in)
FEED/MAGAZINE
Bolt action; 5-round internal
magazine
RANGE
365.76m (400yds)

ARISAKA 38TH
COUNTRY OF ORIGIN
Japan
DATE
1905
CALIBRE
6.5mm (.256in)
WEIGHT
3.95kg (8.7lb)
OVERALL LENGTH
1270mm (50in)
FEED/MAGAZINE
Bolt action; 5-round internal
magazine
RANGE
500m (550yds)

HANYANG 88
COUNTRY OF ORIGIN
China
DATE
1895
CALIBRE
7.92mm (.312in)
WEIGHT
4.08kg (9lb)
OVERALL LENGTH
1110mm (43.7in)
FEED/MAGAZINE
Bolt action; 5-round box
magazine
RANGE
500m (550yds)

bayonet were produced as well. A small number of the Type 35 Naval Rifle variant, with modified sights, were also completed.

Combat experience during the Russo–Japanese War laid bare the shortcomings of the Type 30 rifle. Among these were the tendency of cartridges to burst in the chamber, maiming the operator, jamming, an arduous cleaning process and a lock that had been ill conceived and allowed powder residue to accumulate, sometimes resulting in flash burns to the faces of unfortunate soldiers.

By this time, Arisaka had been elevated to the rank of general, and he teamed with prolific firearms designer Kijiro Nambu, himself an officer in the Japanese Army, to produce the 38th Year rifle. Production of the 38th year, or Type 38, rifle began in 1906 and it remained in service through the end of World War II. The Type 38 fired the Arisaka 6.5mm (.256in) cartridge, which produced little recoil and was underpowered by many Western standards, while its length of 1270mm (50in) challenged the dexterity of the average Japanese infantryman, who stood only 1.6m (5ft 3in) tall. Still, the rifle was said to have been capable of a high rate of fire up to 30 rounds a minute from a five-round box magazine.

Replacing the Type 38
When the Japanese initiated a program to replace the Type 38 with the new Type 99 rifle in 1939, progress was interrupted by the outbreak of World War II. More than 3.4 million of the bolt action Type 38 were completed at five manufacturing facilities by 1940, and the weapon was prominent on all fronts for the Japanese during the Pacific War. A smaller Type 38 carbine was introduced and this was followed by the Type 44 Carbine in 1911. The

Type 97 sniper rifle was adapted for service in the mid-1930s.

Both Japanese and Chinese service rifles of the early twentieth century were influenced by the German Mauser design. The Chinese Gewehr 88 rifle was in use during the Second Sino–Japanese War, which began in 1937. Those Gewehr 88 rifles in Chinese service that were domestically manufactured were often referred to as the Hanyang 88 in reference to the Hanyang Arsenal near the city of Wuhan. A virtual copy of the German Gewehr 88, the Hanyang 88 was chambered for the 7.92mm (.312in) Mauser round, and it was loaded with a five-round clip that fed an external box magazine.

The Hanyang 88 did feature one major advantage over the Japanese Arisaka series of combat rifles. It fired the Mauser round that was significantly heavier than the 6.5mm (.256in) Japanese cartridge. Therefore, the Chinese sometimes enjoyed a slight firepower advantage during close-quarter infantry firefights. The Hanyang 88 was relatively inexpensive to

BERGMANN MP 18
COUNTRY OF ORIGIN
Germany
DATE
1918
CALIBRE
9mm (.35in) Parabellum
WEIGHT
4.2kg (9.25lb)
OVERALL LENGTH
815mm (32in)
FEED/MAGAZINE
Blowback; 32-round
detachable drum magazine
RANGE
70m (76.55yds)

produce, and China's weak economy and industrial base were able to support the demand for the rifle reasonably well.

By the 1930s, the Hanyang 88 had been in service approximately 40 years. It remained in production until 1947 and over 1.1 million were built.

Submachine Guns

Weapons technology advanced rapidly during the years preceding World War I, and automatic weapons were beginning to find greater acceptance as the conflict dragged on. Considered the first blowback-operated submachine

RIGHT: German soldiers rush down a trench in preparation for an assault along the Western Front. Submachine guns began to appear on the battlefield late in the war; however, their numbers were few.

gun in the world, the Bergmann MP 18 had reached the Western Front in limited numbers by 1918.

German weapons engineer Hugo Schmeisser began development of the Bergmann MP 18 in 1916, and the innovative weapon took the name of Theodor Bergmann, whose manufacturing concern, Theodor Bergmann Waffenbau Abteilung, initiated production two years later. The MP 18 was issued to German troops in a matter of months, and its firepower was highly effective in providing suppressive fire support and for clearing enemy trenches.

Full-scale production of the MP 18 began in early 1918, and while exact production numbers remain elusive, historians believe that between 5000 and 10,000 were completed by the end of World War I. During Germany's last, futile attempt at favourable decisive action in the West, the Michael Offensive in the spring of 1918, the presence of the MP 18 in the hands of newly formed Stormtrooper units startled Allied troops.

With its high degree of mobility, the MP 18 was in service during the

MP 28
COUNTRY OF ORIGIN
Germany
DATE
1928
CALIBRE
9mm (.35in) Parabellum
WEIGHT
4kg (8.82lb)
OVERALL LENGTH
820mm (32.28in)
FEED/MAGAZINE
Blowback; 32-round detachable magazine
RANGE
150m (164.04yds)

PISTOLE MIT V-PEROSA M1915
COUNTRY OF ORIGIN
Italy
DATE
1914
CALIBRE
9mm (.35in)
WEIGHT
3.67kg (8.09lb)
OVERALL LENGTH
900mm (35.43in)
FEED/MAGAZINE
Blowback; 25-round box magazine
RANGE
1800m (2200yds)

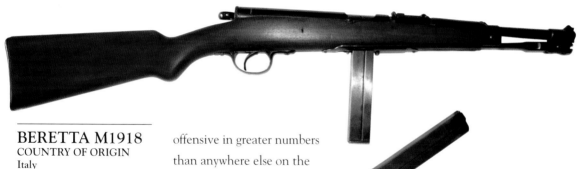

BERETTA M1918
COUNTRY OF ORIGIN
Italy
DATE
1918
CALIBRE
9mm (.35in)
WEIGHT
3.3kg (7.3lb)
OVERALL LENGTH
1092mm (43in)
FEED/MAGAZINE
Blowback; 25-round
detachable box magazine
RANGE
100m (109.36yds)

offensive in greater numbers than anywhere else on the Western Front, and its distinctive report and rate of fire up to 500 rounds a minute were comparable to that of less mobile machine guns. The weapon was fed by a 32-round detachable drum, and it fired the 9mm (.35in) Parabellum round. The MP 18 was followed by the MP 28, which was virtually the same weapon with the introduction of a fire mode selector and tangent sight.

Although the MP 18 was not technically the world's first submachine gun, it is considered the first serviceable weapon of its kind for infantry operations. It was preceded by the Italian Villar-Perosa in 1915; however, the Villar-Perosa was initially intended for combat use by Allied aircraft and was later adapted to infantry support. The Treaty of Versailles outlawed the manufacture of the Bergmann MP 18; however, the Germans were undeterred and clandestine production continued into the 1920s.

While the Bergmann MP 18 presented advanced technology, its appearance late in the war and in too few numbers failed to influence the outcome of the conflict. However, the viability of the weapon was proven in combat, and it initiated the subsequent development of submachine guns and individual automatic weapons.

Some firearms experts assert that the Villar-Perosa M1915 and another Italian weapon, the OVP M1918, could be described as variants of the first deployed submachine gun. Following the introduction of the Model 1915 as an aircraft weapon, it was modified to provide mobile fire support for alpine troops. Its twin barrels provided a substantial rate of fire but the weapon was extremely inaccurate. Despite the fact that it was mobile, the Model 1915 was somewhat unwieldy and was fired from a bipod or a platform worn over the shoulders of the soldier in 'cigarette girl' fashion. The OVP Model 1918 utilized a rifle stock and was basically a half-sized version of the Model 1915 that was capable of single shot mode.

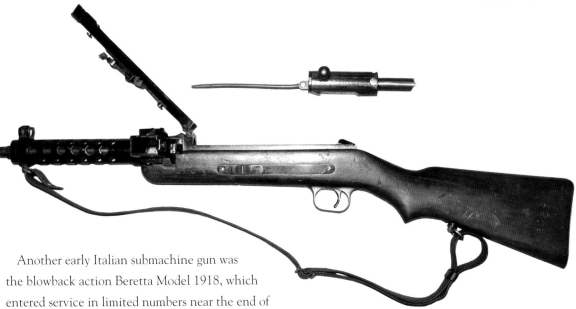

Another early Italian submachine gun was
the blowback action Beretta Model 1918, which
entered service in limited numbers near the end of
World War I. The Beretta fired a 9mm (.35in) Glisenti round and
was fed by a 25-round detachable box magazine that was inserted from
the top and was capable of automatic mode only. A second version, the
Model 1918/30 was loaded from the bottom and equipped with a bayonet.

Treaty of Versailles

Although the German armed forces were prohibited after World War I
by the Treaty of Versailles from deploying most automatic weapons, they
continued to develop them during the interwar years. Their value on the
battlefield had been recognized during the waning days of the Great War and
was further validated by German soldiers that had been deployed to assist
the Nationalist forces of Generalissimo Francisco Franco in his seizure of
power during the Spanish Civil War of the mid-1930s.

German submachine guns usually carried the designation 'MP' for
maschinenpistole, and in its continuing effort to circumvent the terms of
the Treaty of Versailles, the German government engineered the acquisition
of the Swiss firm of Solothurn by domestic arms manufacturer Rheinmetall.
Soon afterward, Rheinmetall purchased a significant ownership stake in the
Austrian arms manufacturer Steyr. The result of multiple collaborative efforts
was a prototype submachine gun that came to be known as the MP 34.

The MP 34 proved to be a highly successful submachine gun, and its
precision construction earned a reputation for operational excellence.
The Austrian version of the MP 34 was known as the Steyr-Solothurn S1-
100. The blowback-operated weapon fired a 9mm (.35in) cartridge that

MP 34
COUNTRY OF ORIGIN
Germany
DATE
1930
CALIBRE
9mm (.35in) Parabellum
WEIGHT
4.25kg (9.4lb)
OVERALL LENGTH
850mm (33.5in)
FEED/MAGAZINE
Blowback; 32-round
detachable box magazine
RANGE
200m (220yds)

ERMA MPE
COUNTRY OF ORIGIN
Germany
DATE
1930
CALIBRE
9mm (.35in) Parabellum
WEIGHT
4.15kg (9.13lb)
OVERALL LENGTH
902mm (35.5in)
FEED/MAGAZINE
Blowback; 32-round box
magazine
RANGE
70m (76.55yds)

STAR SI 35
COUNTRY OF ORIGIN
Spain
DATE:
1935
CALIBRE
9mm (.35in)
WEIGHT:
3.74kg (8.25lb)
OVERALL LENGTH
900mm (35.45in)
FEED/MAGAZINE
Blowback; 40-round
detachable box magazine
RANGE
50m (54.68yds)

was fed by a detachable box magazine holding from 20 to 32 rounds. The weapon offered the operator a selective shot option, either single round or fully automatic. The MP 34 or Steyr-Solothurn S1-100, depending on its origin, was a precision-crafted weapon, and even though it was well made its prohibitive cost limited production.

Another German submachine gun, the Erma MPE, entered production in 1930 and was widely deployed during the Spanish Civil War. Designed in the 1920s by Heinrich Vollmer, the Erma MPE was used by both French Foreign Legion and German troops during the 1930s and was fed by a 32-round detachable box magazine that was loaded from the left-hand side. The weapon featured a wooden stock and pistol grip for control and was later exported in large numbers to several nations in Central and South America whose governments were right wing.

Other European Submachine Guns
Other European submachine guns of the World War I period and the interwar years included the Star SI 35 that originated in Spain. Its complicated design prevented the Spanish automatic weapon from reaching troops in large numbers during the country's civil war, and it was eventually abandoned in favour of cheaper weapons. Its primary characteristics included a single-shot or automatic firing options and easy reloading with the engagement of a device that held the bolt open when the magazine was

empty. The Czechoslovakian-made ZK 383 submachine gun was produced primarily for export beginning in 1938 and fired a 9mm (.35in) Parabellum round compatible with German weapons of the period. The Czech gun fired up to 700 rounds a minute from a detachable box magazine that held up to 40 rounds. After the German occupation of Czechoslovakia many of the weapons were issued to the Waffen SS.

Firepower at the squad level among U.S. Army troops during World War I was augmented by the Browning Automatic Rifle, designed by prolific American firearms engineer John Browning. The Browning Automatic Rifle, or BAR, was technically classified as a rifle but exhibited numerous characteristics of a light machinegun or portable automatic weapon that could be carried by a single soldier. The BAR was chambered to fire the standard American Springfield .30-06 rifle cartridge and was supplied by a 20-round detachable magazine.

Following World War I, the BAR was fitted with a bipod and designated the M1918A1. A slightly improved variant, the M1918A2, was subsequently introduced. Depending on which variant was deployed, the BAR was capable of firing 350 or 550 rounds a minute in fully automatic mode. The gas-operated weapon was also capable of firing single shots. Despite its weight of nearly 7.25kg (16lb) and its rather paltry 20-round magazine, the BAR was popular with American fighting men during both world wars. The bipod featured in later variants was often discarded to reduce the weight of the weapon. Approximately 100,000 of the original M1918 were manufactured.

ZK 383
COUNTRY OF ORIGIN
Czechoslovakia
DATE
1938
CALIBRE
9mm (.35in) Parabellum
WEIGHT
4.83kg (10.65lb)
OVERALL LENGTH
875mm (34.45in)
FEED/MAGAZINE
Blowback; 30-round box magazine
RANGE
100m (109.36yds)

BROWNING M1918
COUNTRY OF ORIGIN
United States
DATE
1938
CALIBRE
7.62mm (.30in)
WEIGHT
7.25kg (16lb)
OVERALL LENGTH
1215mm (47.8in)
FEED/MAGAZINE
Gas-operated tilting breech block; 20-round straight box magazine
RANGE
1500m (1640.42yds)

World War II

When the armoured and infantry spearheads of the German Wehrmacht rolled across the frontier with neighbouring Poland on 1 September 1939, igniting World War II in Europe, the Imperial Japanese Army had been fighting a war of conquest on the Asian mainland for eight years. The instruments of totalitarian political and territorial ambition in a more immediate and deadly form, these armies – along with their eventual Axis partners – brought about the largest and costliest conflict in human history.

By the middle of the twentieth century, some weapons of land warfare had continued to develop beyond the capabilities of the generation of shoulder arms that were utilized during the Great War that had begun a mere 25 years earlier. However, many common bolt-action rifles and rudimentary automatic weapons continued in service for at least one of two reasons – either they were readily available in adequate numbers or their superior performance made their extended service lives an economic and military certainty.

Simultaneously, the living laboratory of a global war brought innovation to

LEFT: An American soldier showing a collection of captured German weapons to a photographer. Among the most prominent is an MP 40 submachine gun.

FUSIL MAS 36
COUNTRY OF ORIGIN
France
DATE
1936
CALIBRE
7.5mm (.295in)
WEIGHT
3.71kg (8.2lb)
OVERALL LENGTH
1021mm (40.2in)
FEED/MAGAZINE
Bolt action; 5-round box magazine
RANGE
365.76m (400yds)

ABOVE: **German troops invade Poland on 1 September, 1939. These soldiers have taken control of a checkpoint on the Polish frontier during the lightning assault that began World War II in Europe.**

the forefront, spawning a legion of deadly battlefield tools that included the semiautomatic, automatic and assault rifles. These weapons multiplied the firepower and killing capacity of the individual soldier many times, ushering in a new era of firearms development and deployment.

Western Bolt-Action Rifles

In 1940, the French Army was the largest land force in Western Europe. Nevertheless, it was by no means the best equipped. The French development, adoption and deployment of modern infantry rifles lagged behind that of other nations. With the outbreak of World War II, most French soldiers still carried the bolt-action Lebel Berthier rifle of World War I vintage. It had entered service in 1907 and undergone some modification in 1915.

The semiautomatic gas-operated Fusil Automatique Modele 1917 rifle that fired a 8mm (.314in) cartridge was introduced in 1917, but it was unpopular with frontline troops and its production run ended within a year. Only a few of the Fusile Automatique 1917 survived in service beyond the mid-1920s.

In 1936, with the deficiencies of antiquated rifles in French service well documented, plans were made to modernize the standard issue French infantry rifle units with the MAS-36, a bolt-action design that would finally retire the bulk of the Berthier and Lebel weapons. At 1021mm (40.2in), the MAS-36 was somewhat shorter than contemporary infantry rifles, and its stocky appearance resembled some carbines of the day. The rifle fired a 7.5mm (.295in) rimless cartridge and was loaded from a clip into a five-round internal box magazine. One of its distinguishing features was a slightly odd forward bend in the bolt to facilitate the grip of the soldier firing the weapon.

Although it represented a substantial improvement over older French rifles, budget constraints resulted in a relative few of the MAS-36 being manufactured by the government-run Manufacture d'Armes de St. Etienne. These were usually reserved for frontline units, where the Lebel and Berthier series continued to serve long past their prime. One stunning example of the situation is the continuing deployment of the Lebel Modele 1886/93 with grenade launcher sections.

Europe Under Threat

The French were not alone in their slow pace of modernization, and other nations of Western Europe that fell under the heel of the Nazi boot continued to equip their ground troops with antiquated shoulder arms. Belgian infantrymen of World War II typically carried the elderly 7.65mm (.301in) FN Modele 1889 rifle, the somewhat controversial Mauser design that was built under license in Belgium. Norwegian infantry squads were equipped with the old bolt-action Krag-Jorgensen Model 1894 rifle that fired a 6.5mm (.256in) cartridge and was fed by a five-round integral magazine.

The Short Magazine Lee-Enfield Mk III, a proven performer during the Great War, continued to equip many units of the British Army during World War II. The first Lee-Enfield dated to 1895, and variants of it remained the primary shoulder arm of the British military for nearly 60 years with more than 17 million produced. The SMLE fired the 7.7mm (.303in) cartridge and was fed by a 10-round magazine loaded with five-round charger clips. The Mk III was also modified to accept new high velocity spitzer ammunition, while its sights and magazine were improved.

As early as 1939, British soldiers also received the Lee-Enfield Rifle No.4

LEE-ENFIELD NO. 4
COUNTRY OF ORIGIN
United Kingdom
DATE
1939
CALIBRE
7.7mm (.303in)
WEIGHT
4.11kg (9.06lb)
OVERALL LENGTH
1128mm (44.43in)
FEED/MAGAZINE
Bolt action; 10-round box magazine
RANGE
1000m (1093.61yds)

Mk I in response to the need for improvements in rifle performance and utility. The upgraded No.4 Mk I provided a lighter, more durable weapon that was simpler to manufacture than its predecessor, the No.1 Mk VI, of which slightly more than 1000 were built and never moved beyond limited trials. Still, the Lee-Enfield No.4 Mk I was not formally adopted by the British Army until 1941.

The No.4 Mk I was easily distinguished from the ubiquitous Mk III SMLE, as its barrel protruded beyond the end of the stock and was somewhat heavier than that of the Mk III, increasing the overall weight of the rifle. By 1942, the No.4 Mk I design was simplified further with the introduction of an indentation on the bolt track to replace the more complex bolt release catch. This version was produced only in Canada and the United States.

The No.4 Mk I was a highly accurate rifle, and a sniper version of the new Lee-Enfield, designated the No.4 Mk I (T), was also introduced. The No.4 Mk I (T) was configured with the standard issue No.4 rifle outfitted with a wooden cheek rest and sniper sight. Some of the sniper rifles were retooled in Canada, while the British firm of Holland and Holland completed most of the conversions. The sniper variant proved successful and remained in service with British forces up to the 1960s. About 1600 examples of an Australian sniper rifle, the SMLE No.1 MK III HT, were fitted with a heavy target (HT) barrel and Model 1918 telescopic sights.

Jungle Fighting

The jungle fighting in the China–Burma–India Theatre created the need for a lighter and shorter version of the standard Lee-Enfield rifle. Therefore, the No.5 Jungle Carbine was developed in the mid-1940s and reached some units of the British Army fighting the Japanese in 1944. The No.5 was an adaptation of a shortened version of the Lee-Enfield No.4 that was originally intended for airborne troops. It was nearly 100mm (4in) shorter and slightly more than 1kg (2lb) lighter than the No.4. Interestingly, the reduction in

LEE-ENFIELD NO. 5
COUNTRY OF ORIGIN
United Kingdom
DATE
1944
CALIBRE
7.7mm (.303in)
WEIGHT
3.24kg (7.14lb)
OVERALL LENGTH
1000mm (39.37in)
FEED/MAGAZINE
Bolt action; 10-round box magazine
RANGE
1000m (1093.61yds)

weight was accomplished by reengineering the barrel and receiver, reducing the amount of wood in the weapon and drilling out the bolt knob. The shortened and lighter No.5 produced a substantial recoil, and this was lessened with the installation of a rubber butt plate. A flash suppressor helped

ABOVE: **French General Alfonse Georges and British commander Lord Gort review troops of the British Expeditionary Force.**

DE LISLE CARBINE
COUNTRY OF ORIGIN
United Kingdom
DATE
1943
CALIBRE
11.4mm (.45in)
WEIGHT
3.7kg (8.15lb)
OVERALL LENGTH
960mm (37.79in)
FEED/MAGAZINE
Bolt action; 7-round detachable box magazine
RANGE
365m (400yds)

RIGHT: **Inventor John C. Garand (left) discusses the attributes of his M1 Garand rifle, the first semiautomatic rifle to be adopted by the armed forces of a major power, with U.S. Army officers.**

conceal the British soldier from detection by the enemy. The Jungle Carbine was fed by a 10-round magazine loaded from five-round charger clips and capable of firing up to 30 rounds a minute. However, such attributes were not as important to the sniper's or jungle soldier's success as accuracy. Weighed in the balance, the No.5 was found wanting. The Jungle Carbine was plagued by what soldiers called a 'wandering zero' because the weapon could not be sighted and expected to fire accurately multiple times. Despite its shortcomings, more than 300,000 of the No.5 were produced between 1944 and 1947.

Sniper Rifle

The British De Lisle carbine was developed in the early 1940s and entered service in 1943 as a silent sniper or long-distance weapon. Designer William De Lisle based the carbine that bears his name on the iconic Short Magazine Lee-Enfield rifle. Innovations such as replacing the barrel with a modified Thompson submachine gun barrel and using modified magazines from the famed Colt Model 1911 pistol allowed the receiver to accept the readily available 11.4mm- (.45in-) calibre cartridge.

A single shot bolt-action weapon, the De Lisle was extremely quiet. In contrast, other silenced weapons of the time had a considerably more audible report. Its effective firing range was 185m (202yds) and its maximum range was 365m (400yds). The single-shot feature of the De Lisle was considered a distinct advantage over semiautomatic operation in that enemy personnel might be alerted with successive discharges. Only 129 examples of the De Lisle were produced from 1943 to 1945, and the majority of these were delivered to British Commandos.

The standard issue rifle of the U.S. Army in World War II was the M1 Garand, a gas-operated rotating bolt weapon that was officially adopted in 1936. The M1 Garand was actually the first semiautomatic rifle to enter regular service with any army in the world. It was designed by John C. Garand, a

Canadian immigrant who had moved as a child to Connecticut with his family.

Garand had previously developed a light machine gun that was adopted by the U.S. Army and taken a job with the U.S. Bureau of Standards. After World War I, he worked as a consultant with the army's Springfield Arsenal and began evaluating the potential of a semiautomatic rifle design.

By 1934, Garand received a patent on his design following 15 years of research and development in an effort to comply with the specifications of the U.S. Army. Production of the M1 began two years later, and the rifle entered service in 1937. The M1 largely replaced the bolt-action Springfield Model 1903, which remained in service primarily as a sniper rifle. Although it was partially replaced by the select fire M14 by the early 1960s, the M1 remained in service for more than 30 years.

M1 GARAND
COUNTRY OF ORIGIN
United States
DATE
1936
CALIBRE
7.62mm (.30in)
WEIGHT
4.37kg (9.5lb)
OVERALL LENGTH
1103mm (43.5in)
FEED/MAGAZINE
Gas operated; 8-round internal box magazine
RANGE
500m (602yds)

M1 CARBINE
COUNTRY OF ORIGIN
United States
DATE
1943
CALIBRE
7.62mm (.30in)
WEIGHT
2.5kg (5.47lb)
OVERALL LENGTH
905mm (35.7in)
FEED/MAGAZINE
Gas operated; 30-round
detachable box magazine
RANGE
300m (328.08yds)

RIGHT: Armed with
the Mauser K98K rifle,
German soldiers move out
somewhere near the front
lines in the early days of
World War II. The Mauser
design served as the basis
for numerous other rifles.

Approximately 6.5 million M1 Garand rifles were produced through 1963. The weapon fired the 7.62mm (.30in) cartridge and was fed by an eight-round en bloc clip through an internal magazine. It was effective up to 402m (440yds), and a well-trained rifleman was capable of firing 40 to 50 accurate shots a minute. This was by far the highest sustained rate of fire of any standard-issue rifle in World War II.

Two sniper versions, the Garand M1C and M1D, were also produced during World War II but never entered active service – even after the M1C was adopted as the U.S. Army's standard-issue sniper rifle in June 1944, complementing the Springfield M1903A4.

Contrary to popular misconception, the light and versatile M1 carbine was not developed as a smaller variant of the M1 Garand rifle. It was, in most respects, its own design. The semiautomatic carbine was issued to light troops such as airborne, headquarters and quartermaster soldiers, as well as to the crews of vehicles and tanks and some officers. Firing a 7.62mm-(.30in-) calibre carbine round, the M1 was intended to give rear echelon troops a heavier weapon than the standard-issue pistols of the World War II period. It actually shared only one common part, a butt plate screw, with the M1 Garand rifle.

Designed by a trio of U.S. Army engineers during three years of research from 1938 to 1941, the M1 carbine entered service in the summer of 1942 and was in common circulation until the 1970s. It was fed by a 15- or 30-round box magazine. The M2, a select-fire fully-automatic variant, was capable of firing up to 900 rounds a minute.

In 1935, the same year that Adolf Hitler repudiated the Treaty of Versailles and publicly acknowledged the modernizing rearmament of the German Army, the Nazi military machine adopted the most recent rifle in the long Mauser line as its standard shoulder arm. The Mauser Karabiner 98 Kurz, also familiar in its abbreviated nomenclature as the K98 or K98k, was a bolt-action rifle that was chambered for the standard Mauser 7.92mm (.312in) cartridge.

During its production run from 1935 to 1945, over 14.6 million of the K98k entered service. Fed by a five-round stripper clip and an internal magazine, the rifle was developed from the Mauser Gewehr 98 that first appeared with German forces around the turn of the century and served as the primary shoulder arm of the German Army in World War I.

During the years between the wars, several variants of the Gewehr 98 were produced, and the K98k was constructed to incorporate the best elements of these. The K98k was lighter at 3.9kg (8.6lb) and shorter at 1110mm (43.7in) than the original Gewehr 98 and was initially described as a short carbine. The Mauser M 98 bolt-action system of the Gewehr 98 was modified from a straight bolt to a better operating turn down bolt in the K98k. Better placement of the iron sights made the weapon accurate up to 500m (602yds). The average rate of fire for a German soldier was 15 rounds a minute.

MAUSER K98K
COUNTRY OF ORIGIN
Germany
DATE
1935
CALIBRE
7.92mm (.312in)
WEIGHT
3.9kg (8.6lb)
OVERALL LENGTH
1110mm (43.7in)
FEED/MAGAZINE
Bolt action; 5-round internal box magazine
RANGE
500m (602yds)

TOKAREV SVT-40
COUNTRY OF ORIGIN
Soviet Union
DATE
1940
CALIBRE
7.62mm (.30in)
WEIGHT
3.9kg (8.6lb)
OVERALL LENGTH
1226mm (48.27in)
FEED/MAGAZINE
Gas operated; 10-round
detachable box magazine
RANGE
500m (602yds)

GEWEHR 43
COUNTRY OF ORIGIN
Germany
DATE
1943
CALIBRE
8mm (.314in)
WEIGHT
4.4kg (9.7lb)
OVERALL LENGTH
1130mm (44.5in)
FEED/MAGAZINE
Gas operated; 10-round
detachable box magazine
RANGE
500m (602yds)

The K98k developed a reputation for ruggedness and reliability in action. Accessories included mounts for the standard infantry bayonet and a grenade launcher intended for clearing fortified houses or other strongpoints. The governments of Turkey and Czechoslovakia were among those of several countries licensed to manufacture the K98k. The Germans deployed a sniper variant of the K98k on all fronts, and 132,000 of these were produced. The rifle was equipped with telescopic sights and accurate up to 1000m (1094yds).

Gewehr 43

Early German efforts to develop a semiautomatic rifle were less productive than the upgrade of the K98k. Designed by the Walther firm, the gas-operated Gewehr 43 rifle showed some promise despite that fact that it was slow and costly to produce and its predecessor, the Gewehr 41, had been a dismal failure. Reports of the impressive performance of the Soviet Tokarev SVT-40 semiautomatic rifle encountered on the Eastern Front led Walther to evaluate captured examples of the SVT-40. Therefore, the gas operating system of the Gewehr 43 was quite similar to that of the Soviet weapon.

The Gewehr 43 fired a 8mm (.314in) round and was loaded from a 10-round detachable box magazine that was fed from a stripper clip. When it entered production in the autumn of 1943, raw materials were in short supply, and only about 400,000 were ultimately produced by the end of the war in 1945. It was issued to a relative few German troops during the final two years of the conflict. More than 50,000 sniper-configured rifles were manufactured.

Italian conscripts deploying to North Africa. The Italian adventure on the African continent proved to be a disaster.

FUSILE MODELLO 91
COUNTRY OF ORIGIN
Italy
DATE
1891
CALIBRE
6.5mm (.256in)
WEIGHT
3.8kg (8.375lb)
OVERALL LENGTH
1285mm (50.6in)
FEED/MAGAZINE
Bolt action; 6-round box magazine
RANGE
500m (602yds)

MP 40
COUNTRY OF ORIGIN
Germany
DATE
1940
CALIBRE
9mm (.35in) Parabellum
WEIGHT
3.97kg (8.75lb)
OVERALL LENGTH
832mm (32.75in)
FEED/MAGAZINE
Blowback; 32-round detachable box magazine
RANGE
70m (76.55yds)

Variants of the elderly Fucile Modello 1891, also known as the Carcano or Mannlicher-Carcano 1891 infantry rifle, were commonly used by Italian troops during World War II, as they had been during Benito Mussolini's military adventures in Ethiopia and Albania during the 1930s. A veteran of service in World War I, the Modello 1891 had entered service with the Italian Army in 1892. Originally chambered for a 6.5mm (.26in) rimless cartridge and fed by an integral six-round magazine and loaded with an en bloc clip, the rifle had begun to show its age by the 1930s.

Reports of failures and poor performance prompted the Italian government to authorize the production of a new weapon, the Modello 1938, which could fire a more powerful 7.35mm (.29in) cartridge. However, problems with the manufacturing process and the limited number of available weapons brought the improvement program to a screeching halt. The Italian Army was forced to revert to 6.5mm (.256in) ammunition and weapons.

For a time, the Italian Army deployed a shorter version of the Modello 1891. However, by 1941 the longer Carcano M91/41, with adjustable sights, was in use.

Western Submachine Guns
German weapons designers helped pioneer the development of the submachine gun. Early versions appeared during World War I, and during the interwar years the Germans made substantial progress in design, manufacture and deployment. With the outbreak of World War II, the MP 40 followed the MP 36 and MP 38 as one of a long line of submachine

guns that stretched back more than 20 years and which came into being primarily due to the need for more rapid production and to take advantage of stamped parts.

The MP 40 and its close relatives are often erroneously referred to as Schmeissers; however, the inventive Hugo Schmeisser was not directly involved in their development. The MP 40 combined the elements of guns from several different designers, although its principal proponent was Heinrich Vollmer, who had also been instrumental in the development of the outstanding MG 34 machine gun.

During the course of World War II, more than a million examples of the open bolt, blowback MP 40 and related weapons were built. The MP 40 fired the 9mm (.35in) Parabellum cartridge and was capable of a cyclic rate of fire of 550 rounds a minute. Although an accomplished operator could squeeze off single rounds with careful pressure on the trigger, the MP 40 offered only the automatic fire mode. The submachine gun was fed by a 32-round detachable box magazine or a 64-round dual magazine configuration, but the 32-round magazine was prone to misfeeds when the user gripped it too tightly to gain stability while firing the weapon. A forward folding stock allowed easy carry but did not hold up well in the field.

Like its predecessors the MP 36, MP 38 and MP 40, the MP 41 submachine gun was conceived to provide individual soldiers with substantial firepower. The MP 41 was essentially the same weapon as the MP 40 with a wooden stock and selective fire option. It fired the 9mm (.35in) Parabellum cartridge at a rate of up to 550 rounds a minute. One impediment to its wide acceptance was a legal battle over patent rights. Originally intended for tank and armoured vehicle crews and officers of small infantry units, the MP 41 was eventually issued only to SS and police detachments.

Early in 1945, Germany initiated production of the MP 3008 machine pistol primarily to equip Volkssturm, or home guard, units with automatic

MP 41
COUNTRY OF ORIGIN
Germany
DATE
1941
CALIBRE
9mm (.35in) Parabellum
WEIGHT
3.87kg (8.5lb)
OVERALL LENGTH
860mm (33.8in)
FEED/MAGAZINE
Blowback; 32-round detachable box magazine
RANGE
200m (218.72yds)

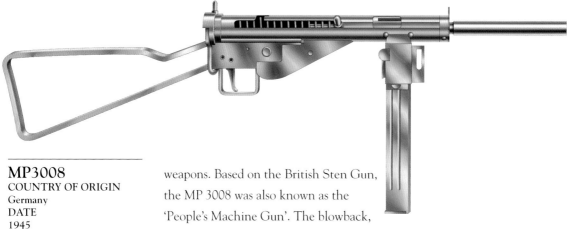

MP3008
COUNTRY OF ORIGIN
Germany
DATE
1945
CALIBRE
9mm (.35in) Parabellum
WEIGHT
3.2kg (7.05lb)
OVERALL LENGTH
760mm (29.9in)
FEED/MAGAZINE
Blowback; 32-round
detachable box magazine
RANGE
70m (76.55yds)

LANCHESTER
COUNTRY OF ORIGIN
United Kingdom
DATE
1941
CALIBRE
9mm (.35in) Parabellum
WEIGHT
4.34kg (9.56lb)
OVERALL LENGTH
850mm (33.5in)
FEED/MAGAZINE
Blowback; 50-round
detachable box magazine
RANGE
70m (76.55yds)

weapons. Based on the British Sten Gun, the MP 3008 was also known as the 'People's Machine Gun'. The blowback, open-bolt submachine gun fired up to 450 rounds of 9mm (.35in) Parabellum ammunition a minute. Cartridges were supplied by a 32-round detachable box magazine. The short MP 3008 production run ended before the war was over, and approximately 10,000 were finished.

Cheap Mass Production

Many of the MP 3008's parts were stamped from steel plate, and the weapon was cheaply produced. Early versions were made without a handgrip, the wire stock was welded to the frame and the finish was rough. When supplies of steel were exhausted, some were finished with wooden stocks.

Acknowledged as a virtual copy of the German Bergmann MP 28, the first British submachine gun of World War II was the 9mm (.35in) Lanchester. With a rate of fire up to 600 rounds a minute, the Lanchester utilized a 32- or 50-round detachable box magazine. Named for its designer, George Lanchester, the weapon entered production in 1941. Although it shared many common parts with the Lee-Enfield rifle, the manufacturing process was complicated and it could not be produced in large numbers. Only about 100,000 were completed by the end of the war.

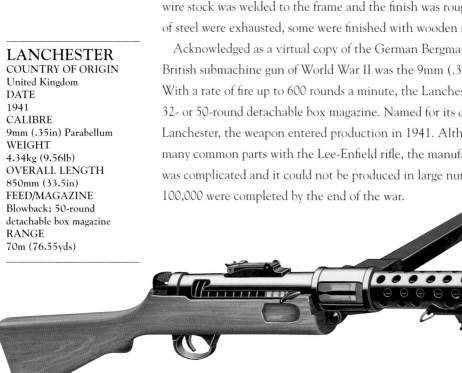

Developed from captured examples of the German MP 40 submachine gun, the famous British Sten is easily recognized. The blowback, open bolt Sten fired a 9mm (.35in) Parabellum cartridge, and its crude metal stock and 32-round box magazine that loaded from the left-hand side were hallmarks of the weapon. The name Sten was derived by incorporating the first letters of the last names of its designers, Major Reginald V. Shepherd and Harold Turpin, along with the first two letters of the Enfield small arms facility where it was produced.

The prototype of the first Sten, designated Mk I, was handmade by Turpin, and about 100,000 were produced with a flash hider, wooden foregrip and forward handle. The later Mk I deleted the wooden parts and the flash hider to speed production.

The only machined parts of the Sten gun were the barrel and the bolt, and the Sten Mk II was cheap to produce as it was made from only 47 stamped and pressed components. More than two million of this mark were completed and widely distributed among Commonwealth forces. The Sten Mk II provided excellent fire support but was prone to jamming, and this tendency was attributed to the expedient production protocol.

The Sten Mk III was the simplest of all the gun's iterations and required only five-man hours to complete its assembly. This version is recognized by a unified receiver, ejection port and barrel shroud that extends further up the barrel than that of the Mk II. Once the barrel was placed inside, the body of the Mk III was welded shut from the top.

Sten Variants

Six British Sten marks were eventually completed and numerous variations were constructed in other countries, while the weapon was widely distributed to resistance networks in Nazi-occupied Europe. The Mk IV was a small Sten that failed to emerge from the prototype stage. The Sten Mk V reached British troops in 1944 and was finished to a higher quality than the

STEN MK I
COUNTRY OF ORIGIN
United Kingdom
DATE
1941
CALIBRE
9mm (.35in) Parabellum
WEIGHT
3.1kg (7lb)
OVERALL LENGTH
760mm (29.9in)
FEED/MAGAZINE
Blowback; 32-round detachable box magazine
RANGE
60m (65.62yds)

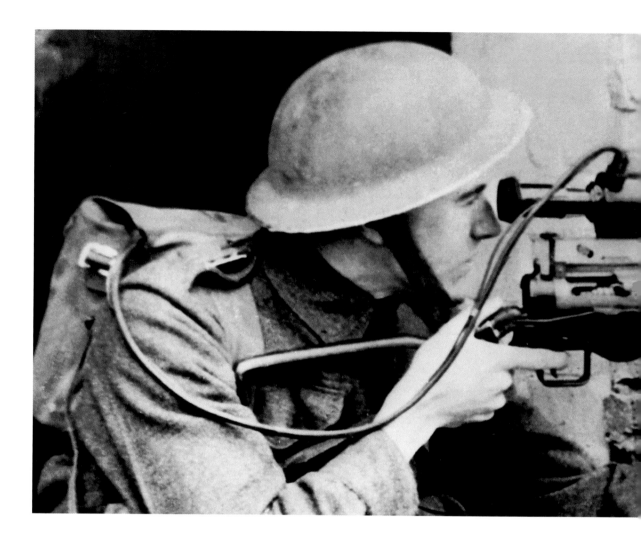

Mk III. The Mk V included improvements such as a wooden pistol grip and
stock. During the decade of the 1940s, more than four million Sten guns
were manufactured.

The British clandestine Special Operations Executive developed a silent
version of the Sten Mk II for use by its agents. The Silent Sten was identical
to the original version with the exception that the muzzle report was
suppressed; however, after firing the first 10 rounds, carbon built up and made
the Silent Sten progressively louder as the 32-round 9mm (.35in) magazine
was emptied. A suppressed version of the Mk V, designated the Mk VI, was
also manufactured.

Australian troops fielded an adaptation of the Sten gun called the Austen,
which was manufactured with several diecasting steps in the process. The
9mm (.35in) Austen entered service with Australian forces in 1942. Slightly

LEFT: A British soldier takes aim through the infrared scope of a Silent Sten Gun. The reduced report of the Silent Sten was developed for use during covert operations.

SILENT STEN
COUNTRY OF ORIGIN
United Kingdom
DATE
1942
CALIBRE
9mm (.35in) Parabellum
WEIGHT
2.95kg (6.5lb)
OVERALL LENGTH
762mm (30in)
FEED/MAGAZINE
blowback; 32-round detachable box magazine
RANGE
70m (76.55yds)

fewer than 20,000 were produced by the end of World War II, and the blowback-operated weapon was capable of firing up to 500 rounds a minute from a Sten-compatible 28-round side-mounted box magazine. An improved model, the M2, and a suppressed variant were introduced later.

The blowback operated Owen gun was the only submachine gun developed in Australia and used in World War II. Designed by

AUSTEN
COUNTRY OF ORIGIN
Australia
DATE
1942
CALIBRE
9mm (.35in) Parabellum
WEIGHT
3.98kg (8.75lb)
OVERALL LENGTH
845mm (33.25in)
FEED/MAGAZINE
Blowback; 28-round
detachable box magazine
RANGE
50m (54.68yds)

PATCHETT MK 1
COUNTRY OF ORIGIN
United Kingdom
DATE
1944
CALIBRE
9mm (.35in) Parabellum
WEIGHT
2.7kg (6lb)
OVERALL LENGTH
685mm (27in)
FEED/MAGAZINE
Blowback; 32-round
detachable box magazine
RANGE
70m (76.55yds)

inventor Evelyn Ernest Owen, it reached the Australian Army in 1943 and fired the 9mm (.35in) Parabellum cartridge fed by a 32-round detachable magazine. About 50,000 Owen guns were produced during the war, and the majority went to troops fighting the Japanese in the jungles of the Southwest Pacific.

By 1944, the British Army had adopted the Patchett Mk I, a submachine gun that, like the Sten, also fired the 9mm (.35in) Parabellum round. The Patchett was intended to supplement the supply of Stens and could accept the detachable magazine of the Sten as well. It could fire at a rate of up to 550 rounds a minute and was constructed through a higher quality process than the Sten. The weapon was named after George Patchett, the chief designer of the Sterling Armaments Company.

Following a rapid series of trials with just over 100 weapons in use, the blowback, open-bolt Patchett was issued to some British airborne troops in time for Operation Market-Garden in the autumn of 1944. The submachine gun was capable of semiautomatic or fully automatic fire and was reasonably accurate and durable in comparison to other British submachine guns of the period, withstanding the rigours of prolonged combat with relatively little maintenance. By 1953, the British Army had officially converted from the Sten to the Patchett, and it was in service until the 1980s.

Thompson Submachine Gun

American inventor John T. Thompson developed the famous Thompson submachine gun in 1919 as a weapon that could possibly break the stalemate of future trench warfare following the heavy bloodshed of World War I. During the Prohibition era of the 1920s and 1930s, the Thompson became a symbol of the ongoing battle between law enforcement officers and organized crime. Both sides used the weapon.

The Thompson submachine gun or 'Tommy Gun' fired a 11.4mm- (.45in-) calibre cartridge and operated with a blowback and Blish Lock breech locking system. Ammunition was supplied by stick or box magazines of 20 or 30 rounds or by 50- or 100-round drum magazines. The Thompson was adopted by the U.S. Army in 1938, nearly two decades after its invention, and provided to British and Commonwealth forces through Lend-Lease during World War II. The first production Thompson was the Model 1921.

The initial version adopted by the U.S. Army was the Model 1928, which differed little from the earlier model but employed a simpler delayed blowback system. The military M1928A1 could accommodate both stick or box magazines. In April 1942, a simplified version of the M1928A1 was adopted by the army and designated the M1. It employed straight blowback action and accepted only the box magazine. More than 1.5 million Thompson submachine guns were manufactured during World War II.

Another American submachine gun, the compact and lightweight Reising, has sometimes been described as a semiautomatic carbine. The delayed blowback, open-bolt Reising appeared in two selective fire variants, the Model 50 and the Model 55 with its folding stock. The Models 60, 65 and a light rifle variant were semiautomatic. The Reising fired a 7.62mm- (.30in-) calibre carbine cartridge and was fed by 12- or 20-round detachable box magazines.

THOMPSON M1921
COUNTRY OF ORIGIN
United States
DATE
1921
CALIBRE
11.4mm (.45in)
WEIGHT
4.88kg (10.75lb)
OVERALL LENGTH
857mm (33.75in)
FEED/MAGAZINE
Blowback; 30-round detachable box or 100-round drum magazine
RANGE
120m (131.23yds)

One of the iconic infantry weapons in the history of the U.S. Army, the Browning Automatic Rifle (BAR) is actually an automatic weapon that almost defies satisfactory classification. Designed by legendary American gunmaker John Moses Browning as a replacement for less desirable French semiautomatic weapons, the BAR was with the American Expeditionary Force in France during World War I.

The BAR established its reputation with the Doughboys and gained lasting fame as a squad-level fire-support weapon during World War II. By 1938, the army had authorized an improvement program to the original BAR, the Model 1918, which had served during World War I. French-designed pistol grips and rate reducer mechanisms were discarded. A rate reducer from the Springfield Arsenal was substituted and provided two separate rates of automatic fire that were chosen by toggle. A bipod was fitted to the muzzle along with a flash suppressor, and adjustable iron sights were installed. The resulting Model 1918A2 was produced in large numbers during the war. It fired the .30-06 rifle cartridge at a rate up to 650 rounds a minute from a 20-round detachable magazine.

Small Arms Pioneer

The BAR is one of many small arms that bear John Browning's name. One of the most significant figures in the development of automatic and semiautomatic weapons, Browning was granted 128 patents during his lifetime. Born in Ogden, Utah, in 1855, he died in Belgium in 1926, already acknowledged as a pioneer and innovator in the small arms industry. Numerous well-known manufacturers produced his designs, including his own Browning Arms Company, Winchester, Remington, Colt, Savage and others.

Among other American submachine guns of World War II were the M3 and its successor, the M3A1. The M3 was popularly known as the 'Grease Gun' or 'Greaser'. The M3 and M3A1 were 11.43mm- (.45in-) calibre

REISING 55
COUNTRY OF ORIGIN
United States
DATE
1941
CALIBRE
11.4mm (.45in)
WEIGHT
2.89kg (6.37lb)
OVERALL LENGTH
787mm (31in)
FEED/MAGAZINE
Blowback; 25-round detachable box magazine
RANGE
120m (131.23yds)

LEFT: Three American soldiers, the leader carrying an automatic weapon, peer cautiously around a corner in Aachen. The seat of government for Charlemagne's Holy Roman Empire, Aachen was the first German city captured by the Allies.

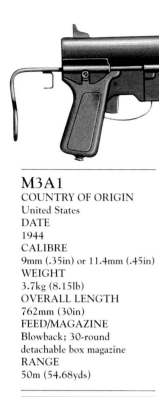

M3A1
COUNTRY OF ORIGIN
United States
DATE
1944
CALIBRE
9mm (.35in) or 11.4mm (.45in)
WEIGHT
3.7kg (8.15lb)
OVERALL LENGTH
762mm (30in)
FEED/MAGAZINE
Blowback; 30-round
detachable box magazine
RANGE
50m (54.68yds)

UNITED DEFENSE M42
COUNTRY OF ORIGIN
United States
DATE
1942
CALIBRE
11.4mm (.45in)
WEIGHT
4.1kg (10lb)
OVERALL LENGTH:
820mm (32.3in)
FEED/MAGAZINE
Blowback; 25-round
detachable box magazine
RANGE
120m (131.23yds)

blowback, open bolt automatic weapons with a rate of fire up to 450 rounds a minute. Production setbacks prevented the M3 from entering service until late 1944. The M3A1 was introduced in December of that year; however, few saw service during World War II. The M3A1 did incorporate several improvements to the original M3, including a more reliable cocking lever assembly. The M3 was prone to fouling from field use, and its single feed, 30-round magazine was difficult to load manually.

The M2 Hyde-Inland had competed with the M3 during trials, and in the spring of 1943 an initial order for more than 160,000 was canceled. Its rate of fire was near 500 rounds a minute, and the M2 shared the same ammunition box magazine feed system as the famed Thompson submachine gun. Only 400 were completed by Marlin Firearms during 1942–43.

The 11.43mm- (.45in-) calibre United Defense M42 was considered as a potential replacement for the Thompson submachine gun. It fired up to 700 rounds a minute and was purchased by the clandestine Office of Strategic Services to supply resistance and covert operations units. Roughly 15,000 were produced from 1942 to 1945.

Along with the Villar-Perosa OVP M1918, Italian troops deployed the Moscheto Auto Beretta MAB 38 submachine gun during World War II. Developed during World War I, the OVP fired a 9mm (.35in) cartridge and was capable of an impressive rate of fire of 900 rounds a minute fed by a top mounted box magazine with a 25-round capacity.

The gas-operated MAB 38 was a 8.8mm (.34in) weapon manufactured at the Beretta Works in Gardone Valtrompia, Brescia. It fired at a maximum of 500 rounds a minute from a detachable box magazine of 10, 20 or 40 rounds. The MAB 38 was capable of semiautomatic or fully automatic firing with two triggers.

Automatic Rifles

The first fully automatic assault rifle to enter combat was the German Sturmgewehr 44, of which 425,000 were manufactured. The bulk of these were sent to the Eastern Front to counter Red Army firepower. Direct interference from Hitler slowed the development of the Sturmgewehr 44, even jeopardizing the weapon's future for a time.

The Sturmgewehr 44 and its predecessor, the Sturmgewehr 43, were essentially identical. The two identifiers were basically subsequent yearly designations for the same weapon. To further complicate matters, the weapon was also known as the MP 43. Although there were minor alterations to the design along the way, the encompassing designation of Sturmgewehr 44, or Storm Rifle 44, became official when the weapon entered full production and is representative of its evolution.

The Sturmgewehr 44 is recognized by its long, curved magazine, which held up to 30 rounds of 7.92mm (.312mm) ammunition, and its sleek lines. It was capable of firing at a cyclical rate of 500 rounds a minute with a range of up to 300m (328yds). The weapon was gas operated with a tilting bolt and capable

STURMGEWEHR 44
COUNTRY OF ORIGIN
Germany
DATE
1944
CALIBRE:
7.92mm (.312in)
WEIGHT
5.1kg (11.24lb)
OVERALL LENGTH
940mm (37in)
FEED/MAGAZINE
Gas operated; 30-round detachable box magazine
RANGE
300m (328yds)

FG 42
COUNTRY OF ORIGIN
Germany
DATE
1942
CALIBRE
7.92mm (.312in)
WEIGHT
4.2kg (9.3lb)
OVERALL LENGTH
945mm (37.2in)
FEED/MAGAZINE
Gas operated; 20-round detachable box magazine
RANGE
500m (602yds)

RIGHT: A weary German soldier stands with his Sturmgewehr 44 automatic rifle slung around his neck. The Sturmgewehr 44 is considered the world's first assault rifle to enter active service.

of semiautomatic or automatic fire. It was designed in the early 1940s, and production began in October 1943. The Sturmgewehr 44 served as the basis for the development of future infantry automatic weapons around the world.

The light Fallschirmjägergewehr 42 (FG 42) was conceived as a weapon that could provide automatic fire support for German airborne troops. Combining the attributes of a machine gun with the easy carry of an infantry rifle, it may be considered an early assault rifle. Designed by Louis Strange and manufactured by Rheinmetall, the FG 42 weighed only 42kg (9.3lb) and fired the 7.92mm (.312in) cartridge from a detachable box magazine of 10 or 20 rounds. A shortage of manganese, used to manufacture the FG 42 barrel, limited production to only 2000 guns. Once the FG 42 reached the field, changes were requested. The bipod was moved from near the hand guard to the muzzle for greater control, the stock construction was changed from metal to wood and the handgrip was slanted to a nearly vertical position.

When the FG 42 was in fully automatic mode, its powerful recoil made the weapon difficult to control. Later versions reduced the rate of fire from 900 rounds a minute to 750, and the muzzle was modified to reduce recoil and muzzle flash.

Charlton Automatic Rifle

The Charlton Automatic Rifle, designed by New Zealand inventor Philip Charlton, was an automatic conversion of surplus Magazine Lee-Enfield and Lee-Metford rifles that dated back to the turn of the twentieth century. The effort was undertaken in part to supplement the meager supply of Bren and Lewis light machine guns available to Commonwealth troops in the Pacific.

Firing the British 7.7mm (.303in) cartridge, the gas-operated Charlton was fed by a 10-round Lee-Enfield magazine or a 30-round magazine that also worked with the Bren Gun. The Charlton could fire at a rate of 600 rounds a minute, and two versions of the rifle were manufactured. The New Zealand version included a bipod and forward pistol grip, while the lighter Australian version did not incorporate either of these.

CHARLTON AUTOMATIC
COUNTRY OF ORIGIN
New Zealand
DATE
1941
CALIBRE
7.7mm (.303in)
WEIGHT
7.3kg (16lb)
OVERALL LENGTH
1150mm (44.5in)
FEED/MAGAZINE
Gas operated; 30-round detachable box magazine
RANGE
910m (995.19yds)

MOSIN-NAGANT 1891

COUNTRY OF ORIGIN
Russia
DATE
1891
CALIBRE
7.62mm (.30in)
WEIGHT
4.37kg (9.625lb)
OVERALL LENGTH
1305mm (51.4in)
FEED/MAGAZINE
Bolt action; 5-round box magazine
RANGE
500m (602yds)

Eastern Bolt-action Rifles

During the opening months of World War II, the primary infantry rifle of the Soviet Red Army was the venerable 7.62mm (.30in) bolt action Mosin-Nagant Model 1891/30. The Model 1891/30 was the product of an effort undertaken in 1924 to modernize the original Model 1891 following the Bolshevik victory in the Russian Civil War.

The Model 1891/30 used the dragoon version of the old Mosin-Nagant as its basis. The barrel was shortened by 89mm (3.5in), and the blade of the front sight was replaced with a hooded post sight. In addition, minor changes were made to the bolt.

In 1936, the Model 1891/30 was again modified, this time for more rapid production. The receiver was simplified from an octagonal to a rounded

shape. With the German invasion of 22 June 1941, production of both the
initial Model 1891/30 and the 1936 variant were increased. By the end of
war, more than 17 million of the basic Model 91/30 had been manufactured.
Although the Mosin-Nagant production rifles of the war years lacked the
overall quality and fine finish of pre-war rifles due to expedited wartime
production, the Model 1891/30 maintained a reputation for rugged
dependability in the often harsh climate of the Eastern Front.

In 1932, the original Mosin-Nagant was modified for use by snipers and
designated the Model 1891/31. With a telescopic sight fitted above the bolt,
the Model 1891/31 incorporated a longer and more curved bolt to allow
smoother operation. The rifle was known for its accuracy, and sniper Vasily
Zaitsev, renowned as a Hero of the Soviet Union and reported to have made
more than 225 kills, made the rifle incredibly famous during the war years.

The Mosin-Nagant Model 1938 Carbine was a shortened carbine version
of the Mosin-Nagant Model 1891/30 rifle and often served as a sniper
weapon. Intended for rear echelon troops, this bolt-action carbine retained
the five-round magazine of the larger rifle. It was designated for replacement
by the Model 1944 Carbine, which was similar to the Model 1938 but added
a spike bayonet. Large quantities of the Model 1944 were produced before
the run ceased in 1948.

Another Soviet carbine, the SKS, was a semiautomatic weapon that
reached Red Army combat troops in 1945. The SKS was configured for
motorized troops that operated in vehicles, headquarters and confined
locations. Its 7.62mm (.30in) ammunition was fed from a 10-round stripper
clip or cartridges could be loaded individually.

LEFT: Clad in winter camouflage, a Soviet sniper peers across a snow covered landscape while holding his Mosin-Nagant rifle equipped with telescopic sight. Red Army snipers claimed many kills with the Mosin-Nagant.

SKS
COUNTRY OF ORIGIN
Soviet Union
DATE
1945
CALIBRE
7.62mm (.30in)
WEIGHT
3.85kg (8.49lb)
OVERALL LENGTH
1021mm (40.2in)
FEED/MAGAZINE
Gas operated; 10-round integral box magazine
RANGE
400m (437.45yds)

When Poland was invaded by the Nazis from the west and the Soviet Red Army from the east, the Polish infantryman defended his homeland with the Karabinek wz. 29, a bolt-action rifle based on the German Mauser-designed Gewehr 1898, the direct forebear of the standard German infantry rifle of World War II, the Mauser K98k. The Karabinek wz. 29 fired the standard Mauser 7.92mm (.312in) cartridge.

The Karabinek utilized a five-round internal box magazine and was fired at an average rate of 15 rounds a minute. Licensed production was undertaken in 1930 at the Polish National Arms Factory in Radom, and 264,000 were manufactured. After the fall of Poland and the partition of

the country, the Karabinek wz. 29 was a common rifle among resistance groups and some organized Polish units that fought from exile with Allied and then Soviet forces.

The vz24 rifle was carried by infantry formations of the Romanian Army during World War II. Manufactured in Czechoslovakia shortly after the end of World War I, the vz24 was designed to closely resemble the German Gewehr 98. It was not considered an identical copy of the German rifle, and a shorter barrel was one of several subtle modifications.

The Imperial Japanese Army of World War II equipped its infantry with two primary bolt-action rifles, the Arisaka Type 38 and Type 99. These were

LEFT: With bayonets fixed to their Arisaka rifles, Japanese soldiers move swiftly toward an objective in the heart of Hong Kong. Fires blaze out of control in the background.

TYPE 99 RIFLE
COUNTRY OF ORIGIN
Japan
DATE
1939
CALIBRE
7.7mm (.303in)
WEIGHT
3.7kg (8.16lb)
OVERALL LENGTH
1120mm (44.1in)
FEED/MAGAZINE
Bolt action; 5-round internal
box magazine
RANGE
500m (602yds)

identified according to the 38th year of the Meiji Dynasty and the year 2099 of the Japanese calendar respectively. Both rifles were heavily influenced by the German Mauser design and were routinely called Arisakas in reference to Colonel Nariakira Arisaka, who led a commission established to develop modern rifles for the Japanese military.

The Type 38 fired a 6.5mm (.256in) cartridge, and battlefield experience in the Sino–Japanese wars of the 1930s indicated the need for a rifle that fired a heavier bullet. Although the Type 99 was intended to replace the Type 38, the advent of war with the United States made total replacement impossible and both rifles served throughout World War II. Nine arsenals produced more than 3.5 million of the Type 99 between 1939 and 1945. Seven of these were located in Japan, while one was at Mukden in occupied China and another was located at Jinsen, Korea.

Ammunition was supplied to the Type 99 by a five-round internal box magazine that was loaded from stripper clips, and the rifle fired a 7.7mm (.303in) cartridge. The Type 99 was also easily recognized with its monopod that was theoretically installed to steady the weapon when firing and for use as an anti-aircraft sight. The rifle was the first to be equipped with a chrome-lined barrel for easier cleaning.

Contrary to some reports of poor quality, the Type 99 rifles built prior to and during the early years of World War II performed well. When Japanese industrial capacity was later crippled and shortages of raw materials occurred, the quality of the rifles diminished rapidly. Each Type 99 was emblazoned on the barrel with a chrysanthemum marking it as the property of Emperor Hirohito.

Two notable variants of the Type 38 rifle, the Type 44 Carbine and the Type 97 sniper rifle, were deployed in World War II. Introduced in 1911, the Type 44 was also referred to as a 'cavalry rifle'. It fired an identical cartridge to that of the Type 38 and was fitted with a needle bayonet. The Type 97 was equipped with a telescopic sight and also fired the small 6.5mm (0.256in) cartridge. It was introduced in 1937 and about 14,000 were made.

During the Sino–Japanese Wars of the 1930s, Chinese forces deployed the Hanyang 88 rifle, a derivative of the old German Gewehr 88. After 1935, another byproduct of the Gewehr 88, the shortened Zhongzheng Type 24 or Chiang Kai-shek rifle, was fielded by both Nationalist and Communist forces fighting the Japanese, and then one another. In the hands of Communist troops, the Chiang Kai-shek rifle was referred to as the Type 79. More than half a million of the weapon were manufactured.

Eastern Automatic Rifles

Feydor Vassilivich Tokarev designed the Samozaryadnaya Vintovka obr 1938, or SVT-38, automatic rifle for the Red Army. The SVT-38 was a long rifle with a simple lock that consisted of a block that cammed downward into the grooved floor of the receiver. When the bolt was moved backward, the lock was released.

The gas-operated SVT-38 was the product of 20 years of research and development by Tokarev. It fired the 7.62mm (.30in) cartridge and was fed by a 10-round box magazine. The SVT-38 was chosen as the new primary Red Army infantry weapon during the mid-1930s, and production began at the Tula Arsenal in the summer of 1939. About 150,000 were manufactured and issued to troops during the Winter War with Finland in 1939–40, but reports that it was cumbersome, difficult to maintain and that the magazine tended to fall out of the weapon precipitated an abrupt suspension of production in 1940.

The lighter and precision-built SVT-40 entered production in July 1940 at facilities in Tula, Izhevsk and Podolsk and incorporated several improvements, particularly a modified magazine release and a hand guard

TYPE 97 SNIPER
COUNTRY OF ORIGIN
Japan
DATE
1937
CALIBRE
6.5mm (.256in)
WEIGHT
3.95kg (8.7lb)
OVERALL LENGTH
1280mm (50in)
FEED/MAGAZINE
Bolt action; 5-round internal box magazine
RANGE
500m (602yds)

TOKAREV SVT-38
COUNTRY OF ORIGIN
Soviet Union
DATE
1938
CALIBRE
7.62mm (.30in)
WEIGHT
3.95kg (8.71lb)
OVERALL LENGTH
1222mm (48.11in)
FEED/MAGAZINE
Gas operated; 10-round detachable box magazine
RANGE
500m (602yds)

RIGHT: Camouflaged Red Army soldiers man a trenchline and await orders from their commanding officer, who scans the horizon. The Soviets supplied large numbers of semiautomatic and fully automatic weapons to their troops.

made of a single piece. It also fired the 7.62mm (.30in) cartridge supplied from a 10-round detachable box magazine. The production process was simplified as well.

Although the SVT-40 was intended initially to replace the bolt action Mosin-Nagant Model 1891-30 rifle, the German invasion of June 1941 necessitated that production of the older rifle be continued. It was later established that one-third of the rifles equipping Red Army divisions should be the semiautomatic SVT-40. However, this ratio was never achieved despite the fact that more than 1.5 million were manufactured during the course of the war.

Variants of the SVT-40 included a carbine and the AVT-40, which was modified with automatic fire capability. The AVT-40, however, was handicapped by its low capacity 10-round magazine, and it also proved somewhat unstable in combat when firing as a fully automatic weapon.

Eastern Submachine Guns

Recognition by the Soviet military establishment of the potential for the submachine gun was a key component in the strength of the Red Army that grew to colossal proportions by the end of World War II. However, the early submachine guns produced in the Soviet Union relied heavily on German designs.

Developed by arms designer Vasily Degtyaryov, the PPD 1934/38 was a virtual copy of the German Bergmann MP 28, and it was fed either by a drum magazine copied from the Finnish Suomi KP-31 or a 25-round box magazine. The PPD 1934/38 was a blowback, open bolt weapon that fired the Soviet 7.62mm (.30in) round. It entered service with the Red Army in 1935 but was found to be too expensive for extended mass production. Although more than 90,000 were manufactured, the more cost effective PPSh-41 was already designated to replace it by the end of 1941.

Developed by designer Georgi Shpagin as a blowback, open-bolt submachine gun that could operate in semiautomatic or automatic firing

TOKAREV AVT-40
COUNTRY OF ORIGIN
Soviet Union
DATE
1940
CALIBRE
7.62mm (.30in)
WEIGHT
3.9kg (8.6lb)
OVERALL LENGTH
1226mm (48.27in)
FEED/MAGAZINE
Gas operated; 10-round detachable box magazine
RANGE
500m (602yds)

PPSH-41
COUNTRY OF ORIGIN
Soviet Union
DATE
1941
CALIBRE
7.62mm (.30in)
WEIGHT
3.64kg (8lb)
OVERALL LENGTH
838mm (33in)
FEED/MAGAZINE
Blowback; 35-round
detachable box or 71-round
drum magazine
RANGE
120m (131.23yds)

SUOMI KP/31
COUNTRY OF ORIGIN
Finland
DATE
1931
CALIBRE
9mm (.35in) Parabellum
WEIGHT
2.8kg (6.17lb)
OVERALL LENGTH
825mm (32.48in)
FEED/MAGAZINE
Blowback; 50-round
detachable box or 71-round
drum magazine
RANGE
70m (76.55yds)

modes, the PPSh-41 was much more cost effective to produce than its predecessor, the PPD 1934/38. During the Winter War of 1939–40 with Finland, senior Red Army commanders learned hard lessons regarding the firepower and mobility of individual soldiers, and in the autumn of 1941 production of the PPSh-41 was undertaken in a number of factories in the Moscow area.

More than 150,000 examples of the PPSh-41 were manufactured during the first five months of 1942, and more than 3000 a day were completed in Soviet facilities as the war progressed. The simple construction of the PPSh-41 included only 87 components and basic stamping and tooling equipment. Amazingly, the machining for a single weapon could be accomplished in little more than seven hours.

By the end of World War II, more than six million of the PPSh-41, with a 71-round drum magazine and a chrome-lined barrel to reduce wear, were manufactured. The Red Army eventually armed entire divisions with it, placing unprecedented firepower in the field.

In early 1943, the PPS-42 submachine gun designed by Alexy Sudayev entered production and proved even less expensive than the PPSh-41 to manufacture. By the end of the year, production had increased to 350,000 weapons a month. The PPS-42 fired the Soviet 7.62mm (.30in) cartridge from a 35-round detachable box magazine and was capable of a rate

of fire of up to 900 rounds a minute. It was primarily issued to support troops, vehicle crews and reconnaissance units. The PPS-43, a slightly modified version of the PPS-42, was designed and produced in Leningrad while the city was enduring a 900-day siege by German and Finnish forces.

Finnish Submachine Guns

The Finnish Suomi KP-31 submachine gun was designed by Aimo J. Lahti and earned an excellent reputation as a serviceable infantry weapon. The original Suomi KP-31 patent was granted in the early 1920s, and later versions of the weapon were rechambered from the 7.62mm (.30in) Soviet

ABOVE: **Outfitted for operations near Sebastopol in the Crimea in 1941, these Soviet partisans are armed with the reliable PPSh-41 submachine gun fed by a 71-round drum magazine.**

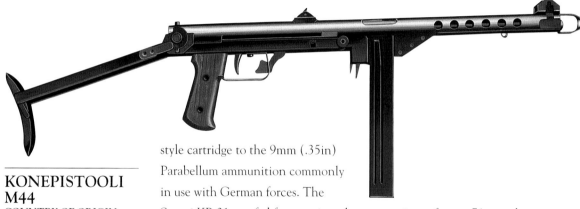

KONEPISTOOLI M44

COUNTRY OF ORIGIN
Finland
DATE
1944
CALIBRE
9mm (.35in) Parabellum
WEIGHT
2.8kg (6.17lb)
OVERALL LENGTH
825mm (32.48in)
FEED/MAGAZINE
Blowback; 50-round detachable box or 71-round drum magazine
RANGE
70m (76.55yds)

BOYS ANTI-TANK RIFLE

COUNTRY OF ORIGIN
Great Britain
DATE
1937
CALIBRE
13.9mm (.55in)
WEIGHT
16kg (35lb)
OVERALL LENGTH
1575mm (62in)
FEED/MAGAZINE
Bolt action; 5-round detachable box magazine
RANGE
90m (98.43yds) against 19mm (.75in) armour

style cartridge to the 9mm (.35in) Parabellum ammunition commonly in use with German forces. The Suomi KP-31 was fed from various drum magazines of up to 71 rounds or staggered row magazines of 20 rounds or more.

The blowback operated KP-31 was capable of selective semiautomatic or fully automatic firing modes. Its remarkable rate of fire exceeded 800 rounds a minute, and during the Winter War with the Soviet Union incidents were documented during which small numbers of Finnish troops held off Red Army forces several times their size with the powerful Suomi KP-31. Undoubtedly, these experiences influenced Soviet military thinking and the deployment of great numbers of submachine guns to Red Army troops during World War II.

Another Finnish submachine gun, the Konepistooli M44, was a copy of the Soviet PPS-43 submachine gun. The M44 was equipped with either the same 71-round drum magazine utilized by the KP-31 or a 50-round magazine. It was later modified for the 36-round magazine of the Swedish Carl Gustav submachine gun of the postwar era.

Japanese Submachine Guns

The only Japanese submachine gun of World War II that was produced in any significant numbers, the Type 100 was inferior to Western counterparts, firing a weak 8mm (.314in) Nambu cartridge and with an initially low

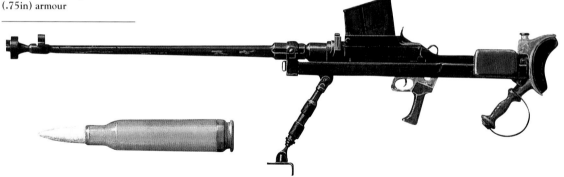

rate of fire of only 450 rounds a minute – although this was later improved to 800 rounds. The open-bolt, blowback Type 100 was fed by a 30-round detachable box magazine mounted on the right-hand side. Based on the German Bergmann MP 18, it entered service in 1942. Japanese industrial capability was limited as the war progressed, and fewer than 30,000 of the Type 100 were completed by 1945.

ABOVE: **A French soldier sights a target on the firing range with the Boys Anti-Tank Rifle. The Boys was a powerful weapon and was well known for its tremendous recoil.**

Anti-tank Rifles

The heavy British Boys anti-tank rifle weighed 16kg (35lb) and was steadied with a bipod and padded butt. A new generation of shoulder-fired anti-tank weapons made the Boys obsolete soon after its debut in 1937. However, approximately 62,000 were manufactured in three basic configurations. The Mk I featured a T-shaped monopod and a circular muzzle brake, while a later Mk I variant had a square muzzle brake and a 'V'-shaped bipod. A smaller airborne variant was manufactured without a muzzle brake and with a shortened 760mm (30in) barrel.

The bolt action Boys was chambered for a large 13.9mm- (.55in-) calibre bullet that was fed from a five-round box clip and capable of penetrating

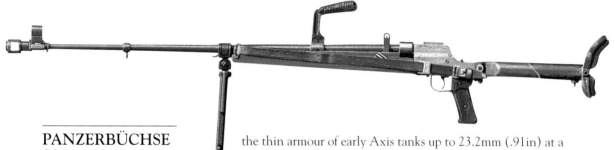

PANZERBÜCHSE 39

COUNTRY OF ORIGIN
Germany
DATE
1939
CALIBRE
7.92mm (.312in)
WEIGHT
11.6kg (25.57lb)
OVERALL LENGTH
1620mm (63.8in)
FEED/MAGAZINE
Bolt action; single shot
RANGE
300m (328.08yds) against
25mm (.98in) armour

GRANAT–BÜCHSE 39

COUNTRY OF ORIGIN
Germany
DATE
1942
CALIBRE
300mm (1.18in)
WEIGHT
10.43kg (23lb)
OVERALL LENGTH
1232mm (48.5in)
FEED/MAGAZINE
Hand feed; single shot
RANGE
125m (136.7yds)

the thin armour of early Axis tanks up to 23.2mm (.91in) at a distance of 91m (100yds). As armour protection increased, the Boys anti-tank rifle was rendered obsolete. However, it was popular for its time despite its tremendous recoil, and it proved effective on the battlefield – Finnish troops deploying the Boys were able to destroy the Soviet T-26 tank during the Winter War of 1939–40.

More than 39,000 Panzerbüchse 39 anti-tank rifles were manufactured for the German Army in 1940–41. Many of these were deployed with German troops during the Battle of France in the spring of 1940, and the single shot 7.92mm (.312in) weapon was reported to have performed well. Designed by engineer B. Brauer and manufactured by Gustloff Werke, the bolt-action weapon was effective against 25mm (.98in) of armour at a distance of 300m (328yds). The introduction of ammunition with a tungsten core increased the penetrating power of the Panzerbüchse round and lengthened the weapon's service life.

In 1942, the German Army deployed the Granatbüchse 39, a variant of its original Panzerbüchse 39 anti-tank rifle. The Granatbüchse 39 was modified to shorten the original barrel of the Panzerbüchse 39, while the bipod, sling band, carrying sling and handle were moved to different locations on the weapon for more efficient carry and battlefield deployment. A threaded

launcher was screwed to the barrel, and the Granatbüchse 39 was capable
of firing three types of grenades – anti-personnel, small anti-tank and large
anti-tank. Each was propelled by a blank 7.92mm (.312in) cartridge.

The Soviet PTRD-41 anti-tank rifle entered service with the Red Army
in 1941 and fired a 14.5mm (.57in) round. The single-shot weapon weighed
a hefty 17.3kg (38.1lb) and required two soldiers to operate. With tungsten
core ammunition it could penetrate up to 40mm (1.57in) of armour at a
distance of slightly more than 100m (109yds). As the armour protection
of German tanks increased, the PTRD-41 was often fired at gun ports or
viewing slits in attempts to disable the vehicle or incapacitate crewmen.
Soviet troops also used the PTRD-41 against entrenched enemy infantry.

Heavier than the PTRD-41 at 20.3kg (46lb), the Soviet PTRS-41 anti-
tank rifle was fed by a five-round magazine and was easier to transport with
a detachable barrel. It fired a powerful 14.5mm (.57in) round with a steel or
tungsten core. Designed by Sergei Gavrilovich Simonov in 1938, the PTRS-
41 was gas-operated and often referred to simply as the Simonov. It was
considered prone to fouling and was not as widely distributed as the PTRD-41.

The Red Army anti-tank rifle squad of 1942 included three teams operating
either the PTRS-41 or the single-shot PTRD-41. The anti-tank platoon
consisted of three squads and totaled nine anti-tank rifles and 24 soldiers.

The Japanese Type 97 anti-tank rifle was extremely heavy at 59kg (130lb)
and packed a tremendous recoil. The gas-operated open bolt semiautomatic
Type 97 fired a 20mm (.79in) round that could penetrate 30mm (1.18in)
of armour at a range of nearly 250m (274yds). It was fed by a seven-round
box magazine. The unwieldy, inaccurate weapon could be disassembled for
transport or hauled with carrying handles, and was operated by a crew of
nine soldiers. Production was commenced in 1938 at the Kokura Arsenal
and terminated in 1941 after 1100 were completed.

TYPE 97 ANTI-TANK RIFLE

COUNTRY OF ORIGIN
Japan
DATE
1937
CALIBRE
20mm (.79in)
WEIGHT
59kg (130lb)
OVERALL LENGTH
2060mm (81.1in)
FEED/MAGAZINE
Gas operated; 7-round
detachable box magazine
RANGE
350m (382.76yds) against
30mm (1.18in) armour; 700m
(765.53 yds) against 20mm
(.79in) armour

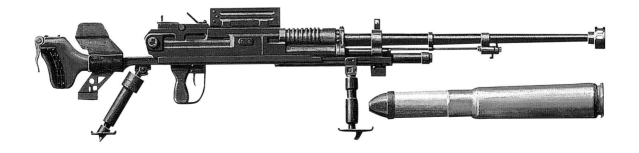

Modern Rifles

Following World War II, the twin phenomena of the Cold War and the burgeoning Third World brought renewed interest in the combat efficiency of the soldier, equipped for the changing modern battlefield with a stunning array of rifles – semiautomatic, automatic, assault – and bolt-action throwbacks still in service as sniper rifles.

The lessons of global conflict were put to use by arms manufacturers old and new, and a proliferation of designs emerged amid the effort to standardize some weapons and ammunition for use by allied countries in the two armed camps of NATO (North Atlantic Treaty Organization) and the Warsaw Pact. At the same time, the rise of nationalism infused with the ardour of political ideology spawned civil wars and hot spots from Central America to Southeast Asia, and from Eastern Europe to the Horn of Africa.

New and deadly rifles emerged, including finely crafted and highly technical implements as well as inexpensive but effective weapons produced by the millions that have become not only the soldier's shoulder arm, but

LEFT: Soldiers take aim with Belgian-designed FN FAL assault rifles. The FN FAL equipped forces of many NATO countries.

FUSIL AUTO MODELE 49
COUNTRY OF ORIGIN
France
DATE
1949
CALIBRE
7.5mm (0.295in)
WEIGHT
4.7kg (10.6lb)
OVERALL LENGTH
1100mm (43.3in)
FEED/MAGAZINE
Gas operated; 10-round detachable box magazine
RANGE
400m (437.45yds)

also the tool of the terrorist and the freedom fighter. Standing above the rest and perhaps the most influential individual weapon of the twentieth century is the ubiquitous AK-47 assault rifle, a symbol of the technology and tumult of the modern era.

Postwar Semiautomatic Rifles

The evolution of the modern rifle continued unabated into the Cold War era, and semiautomatics became common as frontline weapons and as a transitional element in a pronounced movement toward fully automatic and select fire rifles to equip the average soldier. The M1 Garand, standard issue for the U.S. soldier in World War II, had ushered in the infantry level semiautomatic rifle.

One of the earlier postwar semiautomatic rifles that appeared in significant quantities was the French Fusil Auto Modele 49, or MAS-49, a product of the government-owned design and production concern Manufacture d'armes de Saint-Étienne (MAS). Although the French had lagged in the development of modern rifles and their country had fallen to the Nazis in 1940, early research and development into semiautomatic weapons had resulted in the limited manufacture of the MAS-38/39 and contributed to a short postwar run of the related MAS-44.

The MAS-49 was adopted by the French Army in the summer of that year and finally designated as standard issue to replace the collection of odd bolt-action rifles that had equipped the country's armed forces for more than 60 years. The rifle's direct impingement gas operation channeled gas from the fired 7.5mm (.295in) cartridge directly into a cylindrical hollow in front and on top of the bolt carrier to cycle the firing sequence. Although fewer than 21,000 of the original MAS-49 were manufactured, more than 275,000 of the shorter and lighter MAS-49/56 were subsequently produced.

The MAS-49 remained the primary shoulder arm of the French soldier until 1979 when the FAMAS assault rifle was chosen as its replacement. It saw action during the Suez Crisis, the Algerian war of Independence and the Vietnam War. A number of them were produced under contract and delivered to Syria.

The Belgian Fabrique Nationale Model 1949, also known as the FN-49 or SAFN, was reliable and well built but suffered from significant competition and was subsequently marketed to non-aligned countries that were hesitant to accept military aid from either NATO or the Warsaw Pact nations. The

RIGHT: A French soldier carries his MAS-49 rifle over his soldier at Dien Bien Phu, the site of the epic battle in Indochina that sealed the fate of French colonial dominion in Southeast Asia.

FN 1949
COUNTRY OF ORIGIN
Belgium
DATE
1948
CALIBRE
7.62mm (.30in)
WEIGHT
4.31kg (9.8lb)
OVERALL LENGTH
1116mm (43.5in)
FEED/MAGAZINE
Gas operated; 10-round
detachable box magazine
RANGE
400m (437.45yds)

VZ.52
COUNTRY OF ORIGIN
Czechoslovakia
DATE
1952
CALIBRE
7.62mm (.30in)
WEIGHT
4.14kg (9.13lb)
OVERALL LENGTH
1005mm (39.6in)
FEED/MAGAZINE
Gas operated; 10-round
detachable box magazine
RANGE
650m (710.85yds)

first contract for the FN-49 came from Venezuela in the spring of 1948.
These rifles were chambered for the 7.92mm (.312in) Mauser cartridge;
however, subsequent orders from other countries resulted in rifles chambered
for the 7.62x63mm (.30-06in) Springfield, 7.62mm (.30in) NATO and
7.65mm (.301in) Argentine rounds.

The FN-49 utilized a gas-operated tilting bolt action and was typically
fed from a 10- to 20-round box magazine. The rifle was deployed during the
Korean War, unrest in the Congo, the Suez Crisis and in the Falklands War
of 1982. Total production exceeded 175,000.

The gas-operated tilting breechblock Samonabíjecí puška vzor 52, or vz. 52,
semiautomatic rifle was a product of Czechoslovakian design and production
in the wake of World War II. Originally chambered to fire a specialized
7.62x45mm cartridge, it was later retooled to accept the 7.62mm (.30in)
cartridge in use by the Soviet Union. The vz. 52 was designed by Jan and
Jaroslav Kratochvíl and entered service in 1952 after five years of development.
Production ceased in 1959 as the vz. 58 assault rifle was being manufactured.

Unlike most vertical breech locking mechanisms, this rifle's bolt tips
forward rather than to the rear. The trigger assembly strongly resembles
that of the American M1 Garand, and the vz. 52 is capable of firing up to
25 rounds a minute from a 10-round box magazine fed by a clip.

Although the Swedish Army was not a direct belligerent during World
War II, designer Erik Eklund came forward in 1941 with the Ljungman AG-
42 direct impingement gas-operated semiautomatic rifle. The AG-42 fired
a 6.5mm (.256in) cartridge from a 10-round box magazine and remained in
service with the Swedish Army until the 1960s. Initial problems with rusting

gas tubes were corrected in the mid-1950s with the installation of stainless steel tubes, along with modified sights and other improvements.

During the 1950s, rights to the AG-42 machining were sold to Egypt, which produced a variant of the original rifle chambered for the Mauser 7.92mm (.312in) cartridge and known as the Hakim rifle. A carbine version was also manufactured to fire the Soviet 7.62mm (.30in) round, which was named the Rashid carbine.

Postwar Automatic Rifles

The successor to the Belgian FN-49 was the select-fire FN FAL, one of the most widely used rifles of the modern era. Manufactured by the country's Fabrique Nationale, which today is the largest exporter of arms on the European continent, the FN FAL was designed by Dieudonné Saive, who also conceived the FN-49, and Ernest Vervier, and it spawned a line of variants that have been in widespread use for more than half a century.

Most production FN FAL automatic rifles were chambered for the NATO 7.62mm (.30in) cartridge. The gas-operated tilting breechblock weapon was fed by a 20- or 30-round detachable box magazine or a 50-round drum magazine. Its rate of fire was up to 700 rounds a minute, although fully automatic mode produced a substantial recoil and a tendency to climb. It was effective up to 600m (656yds) over an aperture rear sight and post front sight. A 36-year production run ended in 1988 and resulted in the manufacture of more than two million examples of the FN FAL that have been exported or license manufactured in more than 90 countries. The

LJUNGMAN AG-42
COUNTRY OF ORIGIN
Sweden
DATE
1942
CALIBRE
6.5mm (.256in)
WEIGHT
4.71kg (10.38lb)
OVERALL LENGTH
1214mm (47.8in)
FEED/MAGAZINE
Gas operated; 10-round box magazine
RANGE
600m (656.17yds)

FN FAL
COUNTRY OF ORIGIN
Belgium
DATE
1954
CALIBRE
7.62mm (.30in)
WEIGHT
4.3kg (9.48lb)
OVERALL LENGTH
1090mm (43in)
FEED/MAGAZINE
Gas operated; 30-round box or 50-round drum magazine
RANGE
600m (656yds)

ABOVE: Soldiers of the Eastern Caribbean Defence Force move out during Operation Urgent Fury on the island of Grenada, 25 October, 1983. They are armed with the FN FAL automatic rifle.

FN FAL armed the troops of so many NATO countries during the Cold War that it earned the nickname 'right arm of the free world'.

The prototype FN FAL was actually completed in 1946. After evaluation, the British Army tentatively adopted the weapon with modifications. However, this decision was later rescinded due to political wrangling. Nevertheless, most NATO countries were already in the process of adopting the FN FAL.

British Commonwealth forces were issued their derivative of the FN FAL, the L1A1 Self Loading Rifle, in 1954. While many components of the L1A1 and the FN FAL are interchangeable, a few differences included the conversion of the L1A1 metric dimensions to British imperial units,

a unique buttstock and changes to the box magazine. The semiautomatic
L1A1 was later augmented by the fully automatic L2A1 with a rate of fire up
to 700 rounds a minute. The L1A1 served with the British Army until the
1980s, when it was replaced by the L85A1 assault rifle.

Other notable FN FAL variants include the Austrian Sturmgewehr 58
and the Argentine FM FAL produced in the country under license from
1960. The designation FM refers to the Argentine state-run manufacturer
Fabricaciones Militares. Both a standard rifle and a para version with
a folding buttstock have been produced, and these were deployed with
Argentine troops to the Falklands in 1982. The Sturmgewehr 58 was once
the standard rifle of the Austrian Army and built under licence, while
Germany ordered a substantial number of the rifles in the late 1950s and
designated it the G1.

In the mid-1970s, FN research and development produced the 5.56mm
(.22in) gas-operated rotating bolt FN FNC. Initial performance issues due to
rushed production forced the FNC from NATO trials, and it was not until
1989 that the new automatic rifle was formally adopted as standard issue
by the Belgian Army. The FN FNC remains in production and fires up to
675 rounds a minute from a 30-round detachable box magazine.

M14 Rifle

In the United States, the selective-fire 7.62mm (.30in) M14 rifle entered
production in 1959 following a five-year course of research and development.
The M14 served as the standard issue rifle of the U.S. Army until its official

FM FAL ARGENTINE
COUNTRY OF ORIGIN
Argentina
DATE
1960
CALIBRE
7.62mm (.30in)
WEIGHT
4.3kg (9.48lb)
OVERALL LENGTH
1090mm (43in)
FEED/MAGAZINE
Gas operated; 30-round box or
50-round drum magazine
RANGE
600m (656.17yds)

FN FNC
COUNTRY OF ORIGIN
Belgium
DATE
1977
CALIBRE
5.56mm (.22in)
WEIGHT
3.84kg (8.47lb)
OVERALL LENGTH
997mm (39.3in)
FEED/MAGAZINE
Gas operated; 30-round
detachable box magazine
RANGE
400m (437.45yds)

M14

COUNTRY OF ORIGIN
United States
DATE
1957
CALIBRE
7.62mm (.30in)
WEIGHT
4.1kg (9.2lb)
OVERALL LENGTH
1126mm (44.3in)
FEED/MAGAZINE
Gas operated; 20-round
detachable box magazine
RANGE
460m (503.06yds)

BELOW: **The M14 rifle, successor to the M1 Garand, was adopted by the U.S. Army in 1959.**

replacement by the M16 in 1970. Its predecessor, the M1 Garand, lacked a fully automatic option, and this was incorporated in the M14 along with a 20-round detachable box magazine that replaced the latest eight-round clip of the M1. The wooden stock of the M14 included a hinged shoulder rest for use in the prone position, and early rifles were equipped with wooden hand guards.

The M14, originally called the T44, was chosen for production in 1957 following competitive trials. More than 1.5 million M14s were eventually produced by Winchester, the Springfield Arsenal and Harrington & Richards through 1964. The M14E2, later designated the M14A1, was a fully automatic version of the basic rifle that was used as fire support at the squad level.

During the 1950s, the Italian armed forces were equipped with the American-made M1 Garand semiautomatic rifle and also with M1s manufactured under license in the country. By the end of the decade it had become apparent that a selective-fire weapon was needed to keep pace with the modernizing armies of world. Therefore, in 1959 the Beretta BM59

entered production essentially as a retooled version of the M1. The BM59 was rechambered to accept the 7.62mm (.30in) NATO cartridge and fitted with a 20-round detachable box magazine. It served into the 1990s when it was replaced by the Beretta AR70/90 series of assault rifles.

CETME

During the late 1940s, fascist Spain founded CETME (Centro de Estudios Tecnicos de Materiales Especiales) to manufacture small arms for the country's military. The government hired German designers who had worked on semiautomatic weapons for the Third Reich, and in 1964 the M58 rifle was officially adopted. Chambered for the 7.62mm (.30in) NATO cartridge, variants of the delayed blowback action rifle were produced throughout the 1970s. It fired up to 600 rounds a minute from a distinctive, slightly curved 20- or 30-round box magazine.

The firearms link between Germany and Spain was evidenced in the licensing of the CETME rifle to the Germans in 1957 and the use of the CETME M58 as a basis for designs to come from German manufacturer Heckler & Koch. An early Heckler & Koch design, the G3, traced its lineage to the German semiautomatic weapons of the late World War II period. Ironically, the German firm licensed the manufacture of the G3 from a Spanish concern that had hired German engineers.

The Heckler & Koch G3 was developed during the mid-1950s and based on the Spanish incarnation of the Sturmgewehr 45, a late-war German weapon. The roller delayed blowback G3 was chambered for the NATO

BERETTA BM59
COUNTRY OF ORIGIN
Italy
DATE
1959
CALIBRE
7.62mm (.30in)
WEIGHT
4.4kg (9.7lb)
OVERALL LENGTH
1095m (43.1in)
FEED/MAGAZINE
Gas operated; 20-round detachable box magazine
RANGE
460m (503.06yds)

H&K G3
COUNTRY OF ORIGIN
Germany
DATE
1959
CALIBRE
7.62mm (.30in)
WEIGHT
4.1kg (9.04lb)
OVERALL LENGTH
1025mm (40.4in)
FEED/MAGAZINE
Blowback; 20-round detachable box or 50-round drum magazine
RANGE
500m (550yds)

BELOW: **An AK-47 assault rifle being fired. Through more than half a century the rifle and its derivatives have armed military organizations and guerrilla forces around the world.**

7.62mm (.30in) cartridge and fired up to 600 rounds a minute from 20-round detachable or 50-round drum magazines. Variants included the fixed-stock G3A3 and the telescoping wooden stock G3A4. Millions were produced from 1958 to 1997, and the G3 heavily influenced future designs. It spawned a line of assault rifles that were produced in several calibres beginning in the 1960s.

BELOW: **An AK-47 assault rifle being fired. Through more than half a century the rifle and its derivatives have armed military organizations and guerrilla forces around the world.**

Cold War Assault Rifles

The most prolific small arm of the modern era emerged from the long shadow of the Cold War. The brainchild of designer Mikhail Kalashnikov, it is estimated the more than 75 million of the AK-47 assault rifle and its progeny have been produced since 1949. The 7.62mm (.30in) gas-operated rotating bolt selective fire assault rifle has armed the foot soldiers of the Soviet Red Army and allies of the Warsaw Pact in Eastern Europe and client states of the Third World across the globe. During more than 60 years its influence has reached the heights of geopolitical strategy and found its way into the hands of guerrillas, freedom fighters and terrorists.

The reputation of the AK-47 is one of simple construction, toughness and endurance on the battlefield and ease of operation. It has simply become a universal weapon in many ways. Capable of a cyclical rate of fire up to 600 rounds a minute, the versatile weapon is fed by a 30-round detachable box magazine with a distinctive curve that allows a smooth feed of ammunition into the chamber, or 20- and 40-round box magazines. An optional 100-round drum magazine is also available.

Design work on the predecessor of the AK-47, the AK-46, began in the waning days of World War II as the Soviet military establishment sought a low-cost production selective-fire rifle that could rival the innovation found in German automatic weapons such as the Sturmgewehr 44, the world's first true assault rifle to see combat in any significant numbers. The AK-47 is actually a combination of the best attributes of other weapons, including the safety mechanism of the Browning-designed Remington 8, the trigger mechanism and lock features of the M1 Garand and the gas ejection system of the Sturmgewehr 44. During the early phase of production, a machined receiver was substituted for the stamped metal receiver that was prone to failure.

AK-47
COUNTRY OF ORIGIN
Soviet Union
DATE
1947
CALIBRE
7.62mm (.30in)
WEIGHT
3.47kg (7.7lb)
OVERALL LENGTH
880mm (35in)
FEED/MAGAZINE
Gas operated; 30-round box or 100-round drum magazine
RANGE
400m (437.45yds)

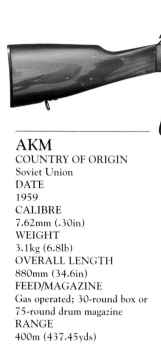

AKM
COUNTRY OF ORIGIN
Soviet Union
DATE
1959
CALIBRE
7.62mm (.30in)
WEIGHT
3.1kg (6.8lb)
OVERALL LENGTH
880mm (34.6in)
FEED/MAGAZINE
Gas operated; 30-round box or
75-round drum magazine
RANGE
400m (437.45yds)

AK-74
COUNTRY OF ORIGIN
Soviet Union
DATE
1974
CALIBRE
5.45mm (.21in)
WEIGHT
3.3kg (7.3lb)
OVERALL LENGTH
943mm (37.1in)
FEED/MAGAZINE
Gas operated; 35-round
detachable box magazine
RANGE
625m (683.51yds)

Over the years, numerous variations on the AK-47 theme have been introduced. They include milled receiver models of 1951 and 1954, and the AKS with a downward folding metal stock rather than the wooden feature so readily associated with the basic weapon. As a testament to the original rifle, more examples of AK-47 variants have been manufactured than any other rifle in history.

AK Variants and Improvements

The best-known variant of the original AK-47 is the AKM, which entered service in 1959. The AKM incorporated the smooth stamped receiver. Alterations were made to the barrel and the gas block, while the bolt carrier is slightly lighter and reshaped. To reduce weight, the wooden stock was extensively hollowed. The basic AKM was extremely portable at only 3.1kg (6.83lb). The AKM is by far the most highly produced version of the early AK-47. Its famous silhouette is recognized the world over.

By the early 1970s, Mikhail Kalashnikov had once again been called upon – this time to develop a possible successor to the AK-47 and AKM. Although the Soviets probably acknowledged that it would never totally replace its iconic predecessors, the AK-74 was widely accepted as an improved member of a growing family of vaunted assault rifles. It entered production in 1974 and remains a stalwart of the Russian and other military organizations around the world. To date, more than five million have been manufactured.

LEFT: A soldier stands on sentry duty with his AK-74 held around his neck by its sling. More than five million examples of the AK-74 have been manufactured since the 1970s.

The resemblance of the AK-74 to its predecessors is unmistakable. However, the rifle does offer some improvements. It is chambered for an intermediate cartridge of 5.45mm (.21in) rather than the older 7.62mm (.30in). A new chrome-lined barrel added service life, while the sight base and gas block were improved. The stock and hand guard were initially manufactured of wood. This was followed by synthetic wood and finally by a hardened dark polymer. The AK-74 was made lighter with cutouts on the stock, and a spring on the lower hand guard tightens the rifle's side-to-side movement considerably during action. Small changes were made to the barrel and 30- or 45-round detachable curved magazines.

The AK-74 first saw action during the Soviet incursion into Afghanistan in 1979 and remains in service today with the armies of most former Soviet republics. Production continues at the Izhevsk manufacturing facility in the Western Urals. Variants include the AKS-74, customized for air assault troops with a metal side-folding stock stamped from sheet metal struts, and the AKSU fully automatic carbine. The RPK-74 standard light machine gun and the RPKS-74 air assault light machine gun with folding stock were, like the AK-74, closely related to the original RPK and RPKS models that were introduced alongside the AKM in the late 1950s.

Among the assault rifles of the Cold War period that borrowed heavily from the AK-47 configuration is the Israeli Galil designed in the l960s after Israeli troops captured large numbers of the AK-47 during the Six Day War of 1967. The gas-operated rotating bolt Galil was chambered for a 5.56mm (.22in) round and capable of firing up to 750 rounds a minute. It entered service in 1972 and remains active today.

In 1980, South African manufacturer Lytton Engineering Works began producing the Vektor R4 assault rifle, a licensed copy of the Israeli Galil. A primary variation in the R4 was the manufacture of

GALIL

COUNTRY OF ORIGIN
Israel
DATE
1972
CALIBRE
5.56mm (.22in)
WEIGHT
3.75kg (8.27lb)
OVERALL LENGTH
850mm (33in)
FEED/MAGAZINE
Gas operated; 65-round detachable box magazine
RANGE
500m (550yds)

the stock and magazine from a strong polymer. About
420,000 of the R4 and its variants have been built. The
Argentine FARA 83 was also heaily influenced by the Galil, entering service
in the mid-1980s with tritium sights for operation in low light and a folding
buttstock. Only 1200 of the FARA 83 were manufactured, and it utilized the
magazine of the Beretta AR70.

A pair of Finnish rifles, the RK-62 and the subsequent Valmet 76 are based
on the AK-47. The Valmet was in production for a decade beginning in 1986
and incorporated a stamped metal receiver rather than the milled receiver of
the RK-62. It saw service during counterinsurgency operations in Indonesia.

Since 1956, China has produced up to 15 million of the Type 56 rifle,
a close facsimile to the AK-47. Early versions of the Type 56 were made
at State Factory 66 with milled receivers. These were later replaced with
stamped receivers, and production was transferred to Norinco (China North
Industries Corporation). It has also been license-made in Bangladesh. The
Type 56 has been widely distributed among the nations of Southeast Asia
and in global hotspots such as the Balkans and Central Asia.

Vz. 58

Despite the historical attention received by the AK-47 and related rifles, the
development and widespread use of assault rifles during the Cold War era
was not the exclusive province of the Soviet Union. The vz. 58, sometimes
erroneously labeled the cz. 58, was of Czechoslovakian origin and replaced
the earlier semiautomatic vz. 52 along with early Soviet-made automatic
rifles in service with the Czech Army.

Although it outwardly resembled the AK-47, its internal workings
differed significantly. For example, the short stroke gas piston had nothing
in common with the Kalashnikov. The extractor and firing pin, both
spring loaded, are inside the falling breechblock, and the fixed ejector is
at the receiver's base. The 7.62mm (.30in) selective fire vz. 58 fired up

TYPE 56
COUNTRY OF ORIGIN
China
DATE
1956
CALIBRE
7.62mm (.30in)
WEIGHT
4.03kg (8.88lb)
OVERALL LENGTH
874mm (34.4in)
FEED/MAGAZINE
Gas operated; 30-round box
magazine
RANGE
400m (437.45yds)

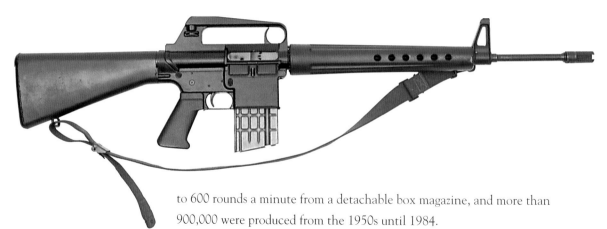

AR-10
COUNTRY OF ORIGIN
United States
DATE
1956
CALIBRE
7.62mm (.30in)
WEIGHT
4.05kg (8.9lb)
OVERALL LENGTH
1050mm (41.3in)
FEED/MAGAZINE
Gas operated; 20-round
detachable box magazine
RANGE
600m (656.17yds)

RIGHT: An American
soldier fires his M16
assault rifle. The M16 is
the military version of the
civilian Armalite AR-15
and was first issued to U.S.
troops in the early 1960s.

to 600 rounds a minute from a detachable box magazine, and more than 900,000 were produced from the 1950s until 1984.

In the United States, the Armalite division of the Fairchild Aircraft Corporation began producing the 7.62mm (.30in) gas-operated rotating bolt AR-10 in 1956. Designed by Eugene Stoner, the AR-10 was distinctive with its straight-line barrel stock and parts made of phenolic resin and forged alloys that dramatically reduced weight. The AR-10 was capable of firing 700 rounds a minute from a detachable 20-round box magazine. In 1957, the basic AR-10 was rechambered to fire the 5.6mm- (.223in-) calibre Remington round and renamed the AR-15. Armalite licensed the production rights for the AR-10 and AR-15 to Colt, and the latter was subsequently adopted by the U.S. Army as its standard-issue assault rifle, the M16.

Rifle, Calibre 5.56mm, M16

The military version of the Armalite AR-15, the M16 assault rifle is forever linked to U.S. involvement in the Vietnam War. It is easily recognized with its top-located carry handle and its prominent triangular front sight. The selective fire gas-operated rotating bolt M-16 was chambered for the 5.56mm (.22in) NATO cartridge and was innovative in its aluminum alloy, composite plastic and polymer construction. Fed by 20- or 30-round box magazines or a 100-round drum magazine, the M-16 can generate a rate of fire up to 950 cyclical rounds a minute.

Design work on the rifle that became the M16 was initiated by Stoner and L. James Sullivan in the mid-1950s. The rifle was deployed with U.S. personnel in Southeast Asia by 1963, and the original finalized production model was designated the M16A1. Variants of the original remain in production, and the rifle is in service with numerous armed forces around the world. To date, more than eight million have been manufactured.

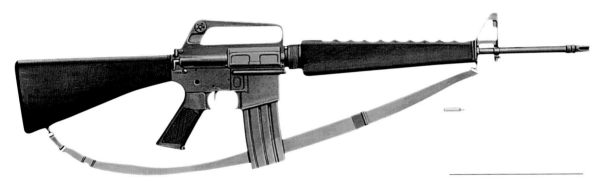

Despite reports that early M16s were prone to jamming due to a phenomenon called 'failure to extract', the service life and overall performance of the assault rifle have been commendable. Ongoing discussion surrounds the negative opinion of some users due to the overall size of the weapon. During desert engagements, the M16 has shown remarkable durability even in conditions with high concentrations of dust and sand.

Variants of the original M16 include the M16A2, introduced in 1983 with a three-round quick burst firing option, the XM177 and M4 Colt Commando carbines, the M16A3, which was primarily issued to U.S. Navy SEAL special forces, the M16A4 that includes a Picatinny rail for the attachment of optics

M16/AR-15
COUNTRY OF ORIGIN
United States
DATE
1962
CALIBRE
5.56mm (.22in)
WEIGHT
3.26kg (7.18lb)
OVERALL LENGTH
1000mm (39.5in)
FEED/MAGAZINE
Gas operate; 30-round box magazine
RANGE
550m (601.49yds)

TYPE 65
COUNTRY OF ORIGIN
Taiwan
DATE
1976
CALIBRE
5.56mm (.22in)
WEIGHT
3.31kg (7.3lb)
OVERALL LENGTH
990mm (38.98in)
FEED/MAGAZINE
Gas operated; 30-round
detachable box magazine
RANGE
500m (550yds)

AR70
COUNTRY OF ORIGIN
Italy
DATE
1972
CALIBRE
5.56mm (.22in)
WEIGHT
3.99kg (8.8lb)
OVERALL LENGTH
998mm (39.3in)
FEED/MAGAZINE
Gas operated; 30-round box or
100-round drum magazine
RANGE
500m (550yds)

and a removable handle, and the heavier barreled Colt Model 655 M16A1 sniper rifle.

When the U.S. Army chose the 5.56mm- (.22in-) calibre M-16, Armalite sought to compete with the Colt licensed assault rifle. By 1963, the AR-18 was chambered for the 5.56mm (.22in) round. Although this rifle was never formally adopted by the military of any country, it did influence further rifle designs. Of particular interest were its stamped steel construction that reduced cost and a gas piston that was significantly more resistant to fouling from residue that the preceding AR-10 and AR-15.

One of the rifles patterned after the AR-18 and its related weapons was the Type 65, developed in the mid-1970s and produced by Nationalist China's Combined Logistics Command. It was formally adopted by the Taiwanese armed forces in 1976, and variants remain in service with police forces today. Another is the Singapore Assault Rifle 80, or SAR 80, adapted for production in Singapore in the late 1970s by American Frank Waters to supplant the AR-15 that had been purchased in the 1960s. On the heels of the SAR 80 came the improved SR 88. Both rifles were chambered for the 5.56mm (.22in) NATO round, but the SR 88 and its follow-on SR 88A introduced higher quality materials into production.

Beretta's AR70 was designed in the early 1970s to replace the aging BM59 and bears some resemblance to the M16 with its carrying handle. By the 1980s, the AR70/90 had become the standard shoulder arm of the Italian Army. The gas-operated assault rifle is chambered for the 5.56mm (.22in)

NATO cartridge and capable
of firing 650 rounds a minute.
It is fed by a 30-round NATO
STANAG magazine or a 100-round drum magazine. The airborne AR90
variant includes a folding stock.

Schweizerische Industrie Gesellschaft (SIG) of Switzerland designed the
SIG SG 510 in the late 1950s, and the heavy 7.5mm (.295in) roller delayed
blowback selective fire rifle remained in service into the 1990s. However, by
the 1970s its potential replacement, the SG 540, was on the drawing board.
At 3.52kg (7.8lb), the SG 540 was significantly lighter than the SG 510, and
a 35-year production run was initiated in 1977. Variants include the SG 542
that was chambered from the NATO 5.56mm (.22in) cartridge to the NATO
7.62mm (.30in) and the SG 543 carbine.

Shoulder Bullpup

Although the bullpup shoulder arm design dates to the turn of the twentieth
century, the concept was reinvigorated with the advent of the Cold War-era
assault rifle. In the bullpup configuration, the buttstock is virtually eliminated
due to the relocation of the action behind the trigger assembly and near the
operator's face. The weight of the firearm is thus reduced, while the barrel
length is unaffected and the rifle is easier to operate and transport.

Among the early Cold War bullpup rifles were the Austrian Steyr-
Mannlicher AUG, widely accepted as the first successful bullpup. With
modular design, optical sights, dual vertical grips and a polymer housing, the
Steyr-Mannlicher AUG was introduced in 1978 and became the primary
rifle of the Austrian and Australian armed forces. The rifle is chambered for
the NATO 5.56mm (.22in) cartridge, while a companion submachine gun,
the Steyr AUG Para, fires the 9mm (.35in) Parabellum round. The weapon
utilizes a 42-round box magazine.

The lever delayed blowback FAMAS assault rifle was introduced in
bullpup configuration in 1981 and followed a succession of French bullpup

STEYR-MANNLICHER AUG
COUNTRY OF ORIGIN
Austria
DATE
1978
CALIBRE
5.56mm (.22in)
WEIGHT
3.6kg (7.9lb)
OVERALL LENGTH
790mm (31.1in)
FEED/MAGAZINE
Gas operated; 30+42-round
box magazine
RANGE
300m (328.08yds)

RIGHT: French soldiers stand at attention with their standard issue FAMAS automatic rifles. The FAMAS was issued to French troops following years of research into bullpup design.

ENFIELD EM-2
COUNTRY OF ORIGIN
United Kingdom
DATE
1951
CALIBRE
7mm (.28in)
WEIGHT
3.49kg (7.7lb)
OVERALL LENGTH
889mm (35in)
FEED/MAGAZINE
Gas operated; 20-round box magazine
RANGE
700m (770yds)

research efforts that dated to the late 1940s. The FAMAS was produced from 1975 until 2000, and more than 400,000 were completed. The standard-issue rifle of the French Army, the FAMAS fires the 5.56mm (.22in) NATO cartridge from a 25- or 30-round box magazine. Introduced in 1994, its G2 upgrade brought the weapon up to NATO standards in several respects. The G2 is capable of firing up to 1100 rounds a minute.

The British Rifle No.9 Mk I, also known as the EM-2 or Janson Rifle, was an early Cold War bullpup designed in the late 1940s by Stefan Kenneth Janson and entering service with British and Canadian forces in 1951. It fired a 7mm (.28in) cartridge and later the 7.62mm (.30in) NATO round from a 20-round box magazine. However, it was not easily adapted to the heavier cartridge and eventually fell from favour.

Later British bullpup production centered on the SA80 (Small Arms for the 80s) group manufactured originally by the Royal Small Arms Factory at Enfield and later BAE Systems. The first of these was the selective fire gas-operated rotating bolt L85A1. Approximately 350,000 of the 5.56mm (.22in) L85A1 and the accompanying L86A1 light support weapon were produced from 1985 to 1994, and the improved L85A2 is now standard issue with the British Army. Three carbine variants of the L85A1 have been produced. Designated the L22 series, the shortened barrel of the carbines has negatively impacted accuracy and power. These have been issued to tank crews and Royal Marines on fleet duty.

In the 1980s, the People's Republic of China introduced the Norinco Type 86 bullpup. The Type 86 was a somewhat reconfigured version of the AKM with a modified selective fire switch, trigger group and vertical folding foregrip. The weapon fires a 7.62mm (.30in) cartridge from a 30-round box magazine and generates up to 600 rounds a minute. Approximately 2000 have been produced.

L85A1 (SA80)
COUNTRY OF ORIGIN
United Kingdom
DATE
1985
CALIBRE
5.56mm (.22in)
WEIGHT
3.82kg (8.4lb)
OVERALL LENGTH
785mm (30.9in)
FEED/MAGAZINE
Gas operated; 30-round detachable box magazine
RANGE
400m (437.45yds)

HK33 Series

When the Heckler & Koch G3 automatic rifle proved commercially successful, the German manufacturer capitalized on its initial effort to produce a family of assault rifles that utilized much of the proven technology of the G3. In a bid to gain a substantial portion of the export market, the HK33 series developed to fire several common cartridges of the day, including the 5.56mm (.22in) and 7.62mm (.30in) NATO, 7.62mm (.30in) M43 Soviet, 7.92mm (.312in) and 9mm (.35in) Parabellum.

The HK33 featured a roller delayed blowback operating system, a conventional hammer-type firing mechanism, rotary rear aperture drum sights, hooded front sights and a selective fire/safety switch. Box magazines of 25, 30 or 40 rounds fed the weapon, and the basic HK33 entered

HK33

COUNTRY OF ORIGIN
Germany
DATE
1968
CALIBRE
5.56mm (.22in)
WEIGHT
3.65kg (8.05lb)
OVERALL LENGTH
920mm (36.2in)
FEED/MAGAZINE
Blowback; 40-round
detachable box magazine
RANGE
400m (437.45yds)

production in 1968. Successive variants are still manufactured today. Some of these include the HK33A2 with a synthetic stock, the HK33A3 with a telescoping metal stock, the HK33 SG1 equipped with a telescopic sight, the HK13 light machine gun with a quick changing heavy barrel and 100-round drum magazine, and the civilian HK C93 semiautomatic produced by Century International Arms.

The Heckler & Koch G41 was produced in Germany from 1984 to 1996 as a possible replacement for the HK33 that was further compliant with NATO standards. The G41 is now licensed to the Italian manufacturer Luigi Franchi.

Cold War Submachine Guns

Postwar Czech arms production included four submachine guns that were designed in the late 1940s, and the most notable of these is the cz. Model 25. One of the Sa series of straight blowback action weapons, the Model 25 used a progressive trigger to move from semiautomatic to automatic firing mode. Chambered to fire the Soviet 7.62mm (.30in) cartridge, it could reach up to 650 rounds a minute. The first variant, the Sa 23 vz. 48a, fired the 9mm (.35in) Parabellum cartridge. The series was the first with a telescoping bolt.

Czech designer Miroslav Rybár began work on the compact cz vz. 61 Skorpion in 1959 for use with special forces and security detachments. However, the 7.65mm (.301in) blowback closed bolt submachine gun was subsequently and somewhat unexpectedly adopted by the Czech Army for issue to lower ranking officers and vehicle crews. The selective fire vz. 61 was later chambered for 9mm (.35in) ammunition in the vz. 65 and vz. 82. The weapon remains in service, and more than 200,000 have been produced.

Concurrent with the Czech foray into submachine guns, the French Manufacture Nationale d'Armes de Tulle (MAT) developed the MAT-49, which entered service in 1949 and was produced for the next 30 years.

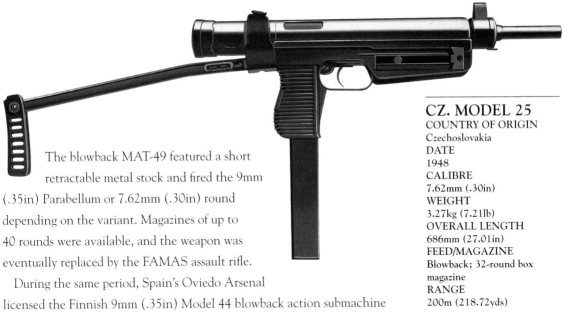

The blowback MAT-49 featured a short retractable metal stock and fired the 9mm (.35in) Parabellum or 7.62mm (.30in) round depending on the variant. Magazines of up to 40 rounds were available, and the weapon was eventually replaced by the FAMAS assault rifle.

During the same period, Spain's Oviedo Arsenal licensed the Finnish 9mm (.35in) Model 44 blowback action submachine gun and produced the DUX-53 and DUX-59 submachine guns primarily for security forces in West Germany. Ironically, the Model 44 had originally been based on the Soviet PPS-43, a successful submachine gun design of World War II.

Argentina introduced the selective fire blowback FMK-3 submachine gun in the mid-1970s, and the weapon saw action during the Falklands War of 1982. The FMK-3 is notable for its centered magazine, blowback open bolt action and 9mm (.35in) Parabellum ammunition. Detachable box magazines up to 40 rounds fed the weapon, and during 1974 – its only year of production – 30,000 were completed.

In 1963, the Lithgow Small Arms Factory of Australia began producing the F1 submachine gun to replace the World War II-vintage Owen Gun and the F88 Austeyr, an Australian licensed version of the Steyr AUG. The 9mm (.35in) Parabellum open bolt F1 featured a fixed firing pin and fired up

CZ. MODEL 25
COUNTRY OF ORIGIN
Czechoslovakia
DATE
1948
CALIBRE
7.62mm (.30in)
WEIGHT
3.27kg (7.21lb)
OVERALL LENGTH
686mm (27.01in)
FEED/MAGAZINE
Blowback; 32-round box magazine
RANGE
200m (218.72yds)

FMK-3
COUNTRY OF ORIGIN
Argentina
DATE
1974
CALIBRE
9mm (.35in) Parabellum
WEIGHT
3.4kg (7.5lb)
OVERALL LENGTH
693mm (27.3in)
FEED/MAGAZINE
Blowback; 40-round detachable box magazine
RANGE
100m (109.36yds)

UZI
COUNTRY OF ORIGIN
Israel
DATE
1948
CALIBRE
9mm (.35in) Parabellum
WEIGHT
3.5kg (7.72lb)
OVERALL LENGTH
640mm (25in)
FEED/MAGAZINE
Blowback; 20, 25 or 32-round detachable box magazine
RANGE
200m (218.72yds)

to 640 rounds a minute from a 34-round box magazine that was compatible with the British Sterling submachine gun, also known as the Patchett, that entered service during World War II. Fewer than 25,000 F1 submachine guns were completed in a short production run during the summer of 1963.

Uziel Gaz Submachine Gun

The design of the legendary Uzi submachine gun is contemporary with the birth of the nation of Israel. In 1948, Israel came into being, and designer Uziel Gal, a major in the Israeli Defence Force, introduced his first open bolt blowback 9mm (.35in) submachine gun. The prototype was finished two years later, and the first Uzis were issued to Israeli Special Forces in 1954.

The Uzi operating system is similar to the Czech Sa 23 series, while it is

constructed primarily from inexpensive stamped sheet metal. Its telescoping bolt allows the stick magazines to fit through the pistol grip, shortening the weapon considerably and making it ideal for close quarter use. With a barrel length of only 254mm (10in), the Uzi achieves a rate of fire of up to 600 rounds a minute, and the precise location of the magazine entry facilitates fast reloading.

Since its inception, more than 90 countries have adopted the Uzi, and it has been manufactured by Israel Military Industries and FN Herstal of France. The Uzi is only 445mm (17.5in) long without its stock, and 635mm (25in) long with the stock extended. Such diminutive stature, along with a light weight of 3.5kg (7.72lb), make it a preferred personal defence weapon. Its design has inspired several other weapons, most of which are referred to as machine pistols. In 1980, the Mini-Uzi was placed in service – its barrel only a scant 19mm (7.76in) long.

Among the Uzi-inspired machine pistols are the Steyr MPi 69 of the 1960s, which shares a telescoping bolt and loading by stick magazine through its grip with the Uzi, and three Spanish machine pistols, the Z-84 introduced in 1985 and the Star Z62 and Star Z70B of the 1960s and 1970s.

Another relative of the Uzi is the American-made Mac-10 blowback machine pistol designed by Gordon B. Ingram in 1964 and produced by the Military Armament Corporation from 1970 to 1973. The primary reason for the straight blowback 11.4mm- (.45in-) calibre Mac-10's notoriety was its two-stage sound suppressor. It fires up to 1145 rounds a minute and remains

HK MP5
COUNTRY OF ORIGIN
Germany
DATE
1966
CALIBRE
9mm (.35in Parabellum)
WEIGHT
2.5kg (5.5lb)
OVERALL LENGTH
680mm (27in)
FEED/MAGAZINE
Blowback; 40-round detachable box magazine
RANGE
200m (218.72yds)

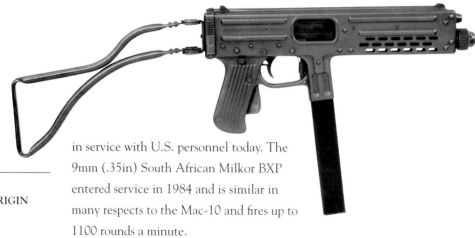

LF57
COUNTRY OF ORIGIN
Italy
DATE
1956
CALIBRE
9mm (.35in) Parabellum
WEIGHT
3.17kg (7lb)
OVERALL LENGTH
686mm (27in)
FEED/MAGAZINE
Blowback; 40-round
detachable box magazine
RANGE
25m (27.34yds)

L42A1
COUNTRY OF ORIGIN
United Kingdom
DATE
1970
CALIBRE
7.62mm (.30in)
WEIGHT
4.4kg (9.7lb)
OVERALL LENGTH
1071mm (42.2in)
FEED/MAGAZINE
Bolt action; 10-round
detachable box magazine
RANGE
730m (800yds)

in service with U.S. personnel today. The 9mm (.35in) South African Milkor BXP entered service in 1984 and is similar in many respects to the Mac-10 and fires up to 1100 rounds a minute.

Heckler & Koch entered the submachine gun market in the mid-1960s with the MP5 (maschinenpistole 5), and over 100 variants of the original 9mm (.35in) roller delayed blowback closed bolt weapon have been produced since then. The selective-fire MP5 is in use with security and military organizations in more than 40 nations, and is one of the most widely deployed weapons of its kind in the world. Variants include the MP5A2 with a fixed butt stock and tricolour firing mode switch, the MP5A3 with a retractable butt stock, the cut-down MP5K and the MP5SD with an integral sound suppressor – popular in the mid-1970s.

Lightweight submachine guns known for personal protection qualities include the 9mm (.35in) blowback Italian LF57 designed by Luigi Franchi in the 1950s. Capable of firing up to 500 rounds a minute, it was limited by low capacity magazines of 25 or 40 rounds. Other such submachine guns include the Italian 9mm (.35in) blowback closed bolt Spectre that debuted in the 1980s and fires up to 850 rounds a minute from a 50-round casket-style magazine, the relatively scarce Finnish 9mm (.35in) Jatimatic of the early 1980s, the Heckler & Koch HK53 KL, a compact version of the HK33K carbine, and the Walther MPK (short) and MPL (long) 9mm (.35in) blowback weapons produced from 1983 to 1985.

Bolt Action and Semiautomatic Sniper Rifles

Although automatic weapons gained momentum steadily in the postwar era, the bolt-action rifle remained viable, particularly when employed in a sniper role. An array of bolt-action sniper rifles continue in use today. Among the best known are the venerable U.S. Springfield M1903A4 and the British L42A1 that served from 1970 to 1990 and was later replaced by the Accuracy International L96. The L42A1, effective up to 730m (800yds), is an updated variant of the Lee-Enfield No.4 Mk I (T) of World War II retooled from the 7.7mm- (.303in-) calibre cartridge to the 7.62mm (.30in) NATO round. A police variant, the Enfield Enforcer, was introduced in the early 1970s.

FR F1
COUNTRY OF ORIGIN
France
DATE
1966
CALIBRE
7.62mm (.30in)
WEIGHT
5.3kg (11.68lb)
OVERALL LENGTH
1200mm (47.24in)
FEED/MAGAZINE
Bolt action; 10-round detachable box magazine
RANGE
800m (874.89yds)

BELOW: The Dragunov sniper rifle was designed as a squad support weapon with long-range capability.

Several other notable sniper rifles of the late twentieth century earned reputations for great accuracy and stability as firing platforms. They include the French FR F1, Mauser SP66, Beretta 501, Steyr SSG 69, FN 30-11, Parker Hale Model 85, SIG SSG-2000 and M40A1.

The FR F1 was manufactured by MAS GIAT Industries and in service from 1966 to 1980. Fitted with a stabilizing bipod and originally chambered for 7.5mm (.295in) ammunition, it was later converted to the NATO 7.62mm (.30in) round. The heavy barreled M40A1 was derived from the Remington Model 700 rifle, chambered for the 7.62mm (.30in) NATO round, and issued in quantity to the U.S. Marine Corps. The Mauser SP66 was adapted from the Mauser Model 66 Super Match Sporting Rifle of the mid-1970s, equipped with a high-powered Zeiss scope and fed 7.62mm (.30in) cartridges from a three-round integral box magazine.

The semiautomatic sniper rifle allows the operator to engage successive targets rapidly. It also dispenses with the time-consuming bolt operation, allowing the sniper to maintain focus and sight picture. Among the world's best of these are the Dragunov, Galil Sniper, PSG1 and MSG90. From Heckler & Koch, the 7.62mm (.30in) PSG1 is based on the G3 selective fire automatic rifle of the 1960s and is said to have been developed for police units following the tragic hostage situation during the 1972 Olympic Games in Munich, Germany. The MSG90 is the military version of the PSG1.

The 7.62mm (.30in) Dragunov entered service in 1963 as a long-range infantry support weapon to offset the loss of distant engagement capability with the common issue of the submachine gun to Soviet infantry units, while the Israeli Galil Sniper is fed 7.62mm (.30in) ammunition from a 25-round box magazine, fitted with a heavy match barrel, and outfitted with a muzzle device that acts as a flash suppressor and muzzle brake.

AN-94
COUNTRY OF ORIGIN
Russia
DATE
1997
CALIBRE
5.45mm (.21in)
WEIGHT
3.85kg (8.49lb)
OVERALL LENGTH
943mm (37.1in)
FEED/MAGAZINE
Gas operated; 45-round detachable box magazine
RANGE
700m (770yds)

The Next Generation

The changing requirements of the battlefield and the continuing search for economic viability in the competitive global arms industry spur everchanging designs and configurations in rifles for military use. Many of these evolve into popular civilian firearms, and the converse is indeed true as fine civilian rifles have been adapted for deployment with military forces.

Traditional assault rifles, compact bullpup and large calibre assault rifles have achieved notoriety as prominent manufacturers seek to satisfy the requirements of diverse constituents. The Soviet-designed AN-94 was once thought to be a contender to replace the AK-74 series with the Red Army; however, its unique two-shot burst feature and limited recoil have failed to offset its cost of production due to complex engineering. Instead, the continuation of the venerable Kalashnikov line, the AK-103, has been in production since the mid-1990s, and more than 200,000 have been built. Other traditional assault rifles include the Heckler & Koch G36, heir to the large following of the successful G3, the 5.56mm (.212in) INSAS rifle that is standard issue with the armed forces of India, and the FX-05 Xiuhcoatl developed in Mexico.

Changing requirements have also resulted in a generation of assault rifles that blazed the trail for the individual weapons of the twenty-first century.

BELOW: A soldier of the U.S. 3rd Infantry Division takes aim on the firing range with his Heckler & Koch G36 assault rifle. Developed in Germany, the G36 is used around the globe.

These include South Korea's K1 submachine gun series undertaken in the 1980s, the Heckler & Koch G11, the Swiss SIG SG 550 and the gas-operated Chinese QBZ-03.

Concurrently, the bullpup configuration has gained momentum with its light weight and compact size. During the 1970s, the Brazilian LAPA FA Modelo 03 was briefly in production, while the Belgian FN F2000 emerged with NATO forces in the 1990s and the Chinese added the QBZ-95 to their arsenal. The Singapore Assault Rifle SAR 21 was introduced in 1999 following four years of design work, and Israel has deployed the modular IMI Tavor TAR 21. The shadowy Iranian military is now carrying the Kyabar KH 2002.

Innovative multicalibre and large calibre assault weapons include the ultimately abandoned Australian Advanced Individual Combat Weapon, the Heckler & Koch XM29 OICW emanating from 1990s design work, the AR-based Sig Sauer 716, the XM25 Counter Defilade Target Engagement System and the FN SCAR.

Sharpshooter Rifle Evolution

The sniper rifle has followed suit with the evolving demands of the battlefield. In the 1960s Heckler & Koch, in collaboration with the Spanish state-owned CETME development firm, developeed the selective fire G3 rifle. Among the numerous variants was the G3SG1 sharpshooter rifle. In the 7.62mm (.30in) G3SG1, certain rifles were gleaned from production and modified with Zeiss optics, an extended stock, a specialized trigger set and other modifications, ushering in a period of exceptional modifications in present and future sniper rifles.

Designed for Swiss security forces in the late 1970s, the 5.56mm (.22in) sniper variant of the SG 550 assault rifle entered service in 1990 and introduced a two-stage trigger, folding stock with adjustable cheek piece,

H&K G3SG1
COUNTRY OF ORIGIN
Germany (Malaysian Variant)
DATE
1959
CALIBRE
7.62mm (.30in)
WEIGHT
4.1kg (9.04lb)
OVERALL LENGTH
1025mm (40.4in)
FEED/MAGAZINE
Blowback; 20-round detachable box or 50-round drum magazine
RANGE
500m (550yds)

ergonomic pistol grip and
stabilizing bipod. Although no
longer in production it remains
in service. American designer
Eugene Stoner debuted the SR-25 Special Application Sniper Rifle in 1990,
a combination of Stoner's iconic AR-10 with the direct gas system of the
AR-15. In 2000, the United States Special Operations Command adopted
the SR-25 because of its faster response time than bolt-action rifles and its
high-capacity 10- or 20-round magazine.

In service with the U.S. Marine Corps since 2008, the M39 Marksman
Rifle is a modified version of the venerable M14 selective fire automatic
rifle. With its pistol grip, stabilizing bipod and adjustable stock, the 7.62mm
(.30in) semiautomatic M39 has primarily been used by a designated
marksman at the squad level when a scout sniper has not been assigned to
the unit. The 7.62mm (.30in) M110 sniper rifle, a potential replacement
for the M39 and the M24 bolt-action system, reached American troops in
Afghanistan in 2010.

A degree of specialization continues to evolve among modern sniper rifles,
and these weapons have often been purpose-designed and manufactured to
fill a specific need. The Accuracy International L115A3 holds the record for
the longest recorded sniper kill as a British soldier dispatched two Taliban
machine gunners at a distance of 2.47km (1.54 miles) with the weapon
that is chambered for the 7.62mm- (.30in-) calibre Winchester or 8.58mm-
(.338in-) calibre Lapua magnum cartridges. The bolt-action L115A3 is a
member of the Arctic Warfare Magnum series of rifles.

M39
COUNTRY OF ORIGIN
United States
DATE
2008
CALIBRE
7.62mm (.30in)
WEIGHT
7.5kg (16.5lb)
OVERALL LENGTH
1120mm (44.2in)
FEED/MAGAZINE
Gas operated; 20-round
detachable box magazine
RANGE
780m (850yds)

L115A3
COUNTRY OF ORIGIN
United Kingdom
DATE
1996
CALIBRE
7.62mm (.30in)
WEIGHT
6.5kg (14.3lb)
OVERALL LENGTH
1200mm (47.2in)
FEED/MAGAZINE
Bolt action; 5-round
detachable box magazine
RANGE
1100m (1203yds)

ABOVE: An American sniper sights a target with his 12.7mm (.50in) Barrett M82A1 antimaterial dual-purpose rifle. The Barrett has become synonymous with excellence in long-range sniper weapons.

The compact 7.62mm (.30in) L129A1 sharpshooter carbine is manufactured by Lewis Machine & Tool in the United States and was deployed with the British military in Afghanistan in 2010. In 1982, the L96A1 was the first sniper rifle in the Arctic Warfare series to be adopted by the British Army.

The 7.62mm (.30in) bolt-action SIG SSG 3000 is a well-known police unit sniper rifle distributed throughout Europe, while the Finnish Sako TRG 22 is an example of a somewhat rare rifle designed as a sniper weapon from inception rather than retooled from an existing automatic rifle design. It is chambered for the 6.6mm- (.26in-) calibre Remington, the 7.62mm (.30in) and 7.82mm (.308in) magnum Winchester, and the 8.58mm (.338in) Lapua magnum rounds.

Heavy Calibre Sniper Rifles

Heavy calibre sniper rifles are capable of tremendous penetrating power at great distances as antipersonnel or antimateriel weapons in disabling light armoured vehicles and other somewhat hard targets. These include the American RAI Model 300 and 500 bolt-action rifles chambered for cartridges up to 12.7mm- (.50in-) calibre and similar to the Soviet Dragunov, and the Hungarian Gepard M1 and M3 models that emerged in the 1990s. Perhaps the most famous antimaterial dual-purpose rifle today is the Barrett M82A1 designed by Ronnie Barrett and developed throughout the 1980s. The 82A1

fires a 12.7mm- (.50in-) calibre cartridge from a 10-round box magazine and has been recognized for its sniper accuracy. Also known as the 'Light Fifty', the 82A1 is equipped with a stabilizing bipod and precision scope. A bullpup configuration version, the 82A2, was produced in limited numbers.

The bolt-action American-made 12.7mm- (.50in-) calibre Mcmillan TAC-50 entered service in 2000 and is the primary long range sniper weapon of the Canadian Army. The single-shot bolt-action 12.7mm- (.50in-) calibre Steyr HS .50 made headlines in 2007 when the Austrian government approved the sale of a quantity of the rifles to Iran, while the British Accuracy International AS50 was designed for British troops and adopted by the U.S. Navy SEALS Special Forces, in part due to its high rate of semiautomatic fire, light weight and ease of disassembly. It can be taken apart for servicing without tools in less than three minutes.

The Harris M87R long-range sniper rifle is a single-shot bolt-action rifle with a five-round magazine, the only substantial difference from the initial M87 model. It is distinguishable by its 736mm (29in) barrel with muzzle brake in place.

The standard long-range sniper rifle of the French Army is the bolt action 12.7mm- (.50in-) calibre Hécate II manufactured by PGM Précision. The Hécate II is equipped with both a front bipod and rear monopod for maximum stability and is often used with high explosive incendiary armour piercing (HEIAP) ammunition to destroy a variety of targets from effective distances of up to 1800m (1970yds). It utilizes a seven-round magazine and has served in Mali and Afghanistan.

MCMILLAN TAC-50

COUNTRY OF ORIGIN
United States
DATE
2000
CALIBRE
12.7mm (.50in)
WEIGHT
11.8kg (26lb)
OVERALL LENGTH
1448mm (57in)
FEED/MAGAZINE
Bolt Action; 5-round detachable box magazine
RANGE
1800m (1970yds)

HARRIS (MCMILLAN) M87R

COUNTRY OF ORIGIN
United States
DATE
1987
CALIBRE
12.7mm (.50in)
WEIGHT
9.53kg (21.01lb)
OVERALL LENGTH
1346mm (52.99in)
FEED/MAGAZINE
Bolt action; 10-round detachable box magazine
RANGE
1500m (1640yds)

Shotguns

The shotgun traces its heritage to the muzzleloading, smoothbore blunderbuss of the seventeenth century. The early blunderbuss was utilized for personal defence, by mail couriers for protection against robbers and in the military. Naval crews were often equipped with the blunderbuss for close quarter combat when boarding ships or staving off an enemy that had come alongside.

Therein lies the essence of the modern shotgun. Typically a smoothbore weapon, the shotgun delivers substantially greater stopping power than the contemporary rifle or handgun. The shotgun most often fires a shell or cartridge filled with small round pellets known as shot, or a dense single projectile referred to as a slug. The low penetrating power of the shotgun

LEFT: The name Winchester, a well-known U.S. manufacturer, has become associated with a variety of shotguns favoured by hunters of small game and with certain military applications.

BLUNDERBUSS
COUNTRY OF ORIGIN
Netherlands
DATE
1600
CALIBRE
17.53mm (.69in)
WEIGHT
4.54kg (10lb)
OVERALL LENGTH
774.7mm (30.5in)
FEED/MAGAZINE
Single shot muzzleloader
RANGE
14m (15.31yds)

RIGHT: Annie Oakley became a legendary expert with firearms as she travelled with Buffalo Bill Cody's Wild West Show and performed trick shots that amazed audiences.

coupled with its high stopping capability offers an advantage in minimizing the risk of collateral damage beyond the intended target.

The shotgun is the preferred weapon for hunting small game, particularly birds, and early examples were regularly referred to as 'fowling pieces'. The shotgun finds other uses in the sporting endeavours of trap, skeet and sporting clays. It is a favourite among law enforcement organizations for riot control and close contact with hostile elements, and it remains in service with military forces as well.

For many years, the classic double barrel break-action shotgun, loaded with two shells by 'breaking' the hinged barrel to load from the breech, has been prominent. Beginning in the late nineteenth century, numerous alternatives have emerged, including pump, lever, semiautomatic and even bolt-action shotguns. The calibre of a shotgun is measured in terms of gauge or bore, and the pattern of shot may be influenced by constriction or widening of the barrel, known as the choke.

Side-by-Side Shotguns

In recent years, the double barrel side-by-side shotgun has become something of a rarity; however, for years it reigned supreme among bird hunters and was referred to simply as the 'double.' One of the most prominent manufacturers of the double barrel shotgun was Parker, founded after the American Civil War. The company manufactured the double barrel shotgun into the 1930s, and one of its best known is the A-1 Special introduced in 1907, typically seen in 12-, 16-, 20- or 28-gauge and richly appointed.

The A.H. Fox HE Grade Super 12-gauge shotgun was produced from 1898 to 1929. Highly prized by collectors today, Fox shotguns were manufactured in several grades from the least expensive AE to the top of the line XE. The Fox brand was reintroduced by Savage in 1939, and multiple shotguns were produced throughout the late 1980s. The L.C. Smith Company produced numerous side-by-side shotguns from the late 1800s until 1950, the year

REMINGTON 1889
COUNTRY OF ORIGIN
United States
DATE
1889
CALIBRE
18.5mm (12g)
WEIGHT
3.6kg (8lb)
OVERALL LENGTH
1000mm (39.37in)
FEED/MAGAZINE
Side by side; double shot
RANGE
20m (22yds)

REMINGTON 1894
COUNTRY OF ORIGIN
United States
DATE
1889
CALIBRE
18.5mm (12g)
WEIGHT
3.6kg (8lb)
OVERALL LENGTH
1000mm (39.37in)
FEED/MAGAZINE
Side by side; double shot
RANGE
20m (22yds)

BROWNING BSS
COUNTRY OF ORIGIN
United States
DATE
1971
CALIBRE
18.5mm (12g)
WEIGHT
3.6kg (8lb)
OVERALL LENGTH
1193.8mm (47in)
FEED/MAGAZINE
Side by side; double shot
RANGE
20m (22yds)

after the company was purchased by Marlin. One of the best-known L.C. Smith shotguns was the breechloading A-3 with automatic ejector and rich embellishment.

A rare find among side-by-side shotguns is the Winchester Model 21, a manual break-action weapon that was produced in 12-, 16-, 20- and 28-gauge, along with .410 bore. The high cost of production during a run from 1931 to 1960 resulted in only about 30,000 being completed.

Among the most recognized side-by-side shotguns manufactured in the late nineteenth century are the Remington Models 1889, 1894 and 1900. The outside hammer Model 1889 was manufactured for nearly 20 years, and before its discontinuation in 1908 approximately 135,000 were produced in 10-, 12- and 16-gauges. Firing via circular action hammers, the Model 1889 was produced in seven grades, from decarbonized steel barrels to superior damascus steel barrels with engraving and a high quality walnut stock.

The break-action Model 1894 was Remington's first hammerless double barrel shotgun, and about 42,000 were produced up to 1910. The Model 1894 was manufactured in 10-, 12- and 16-gauge varieties with an optional automatic ejector and in damascus or ordnance steel barrels. The Model 1900 was developed as an inexpensive version of the Model 1894, and production of the 12- and 16-gauge shotgun approached 100,000 before ceasing in 1910.

In 1878, Daniel LeFever introduced the first hammerless shotgun, and two years later he founded his own company. The LeFever Arms Company did not produce as many doubles as competitors with greater capacity;

however, their quality was superb. LeFever, who produced the first automatic hammerless shotgun in 1883, sold his company to Ithaca in 1916.

Browning did not manufacture a side-by-side shotgun until 1971, and production ceased in 1987. Prominent in the short duration of Browning manufacture was the BSS, manufactured in 12- and 20-gauge with pistol or straight grip stocks and barrel lengths of up to 762mm (30in).

BELOW: Adolf Toepperwein was a noted marksman and firearms collector. He owned many shotguns, including this vintage Winchester held by auctioneer Robb Burley.

BROWNING BPS
COUNTRY OF ORIGIN
United States
DATE
1978
CALIBRE
18.5mm (12g)
WEIGHT:
3.49kg (7.7lb)
OVERALL LENGTH
1219.2mm (48in)
FEED/MAGAZINE
Pump; three-round magazine
RANGE
40.23m (44yds)

BENELLI NOVA
COUNTRY OF ORIGIN
Italy
DATE
1990
CALIBRE
18.5mm (12g)
WEIGHT
3.63kg (8lb)
OVERALL LENGTH
1257.3mm (49.5in)
FEED/MAGAZINE
Pump; 4-round internal
magazine
RANGE
30m (32.81yds)

Pump-action Shotguns

In 1893, prolific gun designer John Browning, then working with Winchester Repeating Arms, introduced the Model 1893, the first pump-action shotgun to reach the market in any quantity. Subsequently, the Model 1897 was adapted for use with smokeless propellant, while the barrel was lengthened, the frame was lengthened and covered to fully eject the spent cartridges from the side and the shotgun was strengthened overall to handle a 12-gauge shell. Fed from a five-round tubular magazine, the Model 1897 was issued to American troops in World War I and earned the nickname of 'Trench Gun'. More than a million were manufactured during 60 years of production.

Today, the Browning BPS pump shotgun is manufactured with a forged and machined steel receiver, bottom ejection and loading for ambidextrous operation and a top tang safety. The BPS is manufactured in 12-, 20- and 28-gauge, and .410 bore. Its three plus one or four plus one magazines are accompanied by a three-shell adapter to comply with hunting laws prevalent in most of the United States. A serrated side release is positioned at the rear of the trigger guard.

The lightweight Nova Pump from Benelli was engineered for durability and performance in the field. The 12-gauge and 20-gauge shotguns include steel skeletal framework covered by a state-of-the-art polymer and a stock and receiver formed as one solid piece to eliminate unnecessary play in the weapon. The synthetic grip surface is grooved, the trigger assembly is removable and the model features a push-button shell stop. A compact 20-gauge model is also made.

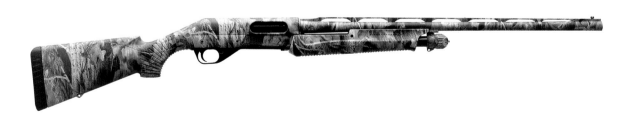

The Winchester Model 12, introduced in 1912, was a standard-bearer for pump-action shotguns during the following half century. Loaded with a six-round tubular magazine, the 12-, 16-, 20- and 28-gauge shotgun was initially available in 20-gauge only. Designed by Thomas C. Johnson, the Model 12 was based on the earlier Models 1893 and 1897 by John Browning. More than two million were made until 1964, and special production runs were conducted until 2006.

In 1915, Browning and cohort John Pederson introduced the design that was to become the Ithaca Model 37, a pump-action shotgun that entered production in 1937 and remains so today. The Ithaca Model 37 is popular with military and law enforcement organizations around the world. An Argentine version is marketed as the Bataan Modelo 71. The Ithaca Model 37 features an ejection and loading port on the bottom of the gun, facilitating use by left- or right-handed operators. It is fed by a tubular magazine of four, five or seven rounds. The Stakeout is a shorter variant with a 330mm (13in) barrel, which has been prominent in numerous motion pictures.

Remington Model 870

The Remington Model 870 is perhaps the best known in a series of pump action shotguns produced by the company. Fourth in the succession, the Model 870 followed Models 10, 17 and 31, and entered production in 1951. The Model 870 remains in production, and 10 million have been manufactured to date. An outstanding hunting, self-defence and sporting shotgun, the Model 870 is also popular with law enforcement agencies.

ITHACA MODEL 37
COUNTRY OF ORIGIN
United States
DATE
1933
CALIBRE
18.5mm (12g)
WEIGHT
3.45kg (7.6lb)
OVERALL LENGTH
1006mm (39.61in)
FEED/MAGAZINE
Pump; 7-round tube magazine
RANGE
30m (32.81yds)

REMINGTON MODEL 870
COUNTRY OF ORIGIN
United States
DATE
1951
CALIBRE
18.5mm (12g)
WEIGHT
3.6kg (8lb)
OVERALL LENGTH
1280mm (50.5in)
FEED/MAGAZINE
Pump; 7-round tube magazine
RANGE
21.95m (24yds)

The genesis of the Model 870 came through competition with the Winchester Model 12, as Remington sought a durable, cost-effective and ergonomic shotgun for the commercial market. The Model 870 is fed by internal tubular magazines in four plus one or seven plus one configurations and in the standard spectrum of 12-, 16-, 20- and 28-gauge, or .410 bore. The immediate predecessor of the Model 870, the Model 31, was designed in 1931 and produced until 1949. Nearly 200,000 examples were completed, and the Model 31 served as a basis for shotgun lines from Maverick and Mossberg, including the Mossberg 500 series.

Mossberg Shotguns

Developed by O.F. Mossberg and marketed by the company that bears his name, the Mossberg 500 series of pump-action shotguns is one of the widest selling in the world. The Model 500 entered production in 1960 and is available in .410 bore and 12- and 20-gauges. Its maximum range is 50m (54.64yds) for shells and 300m (328yds) for slugs. Tubular internal magazines vary from five plus one to eight plus one capacities. Designed for easy maintenance, the Mossberg 500 is simple to disassemble. Numerous variants are produced for military, law enforcement and self-defence. A popular security variant of the Model 500 is the Model 535 introduced in 2005 with both smoothbore and rifled barrels that fire either shells or slugs, while the 835 ULTIMAG is a popular hunting shotgun that accommodates either light or magnum shells.

The Winchester Models 1200 and 1300 were produced from 1965 to 2006, with the Model 1300 taking over from 1983 through to the end of the line. The Model 1200 was originally intended to replace the iconic Model 12. The pump-action shotgun has a five-round capacity with one shell chambered and four in its tubular magazine. The military variant of the Model 1200 is known for its ease of carry when broken down for transportation.

The Model 1300 is slightly modified with a five-round tubular magazine, and the magazine in the Model 1300 Defender variant holds seven rounds.

MOSSBERG 835 ULTIMAG
COUNTRY OF ORIGIN
United States
DATE
1988
CALIBRE
18.5mm (12g)
WEIGHT
3.29kg (7.25lb)
OVERALL LENGTH
1022.35mm (40.25in)
FEED/MAGAZINE
Pump; 3-round magazine
RANGE
50m (54.68yds)

LEFT: The Remington Model 870 is one of the most popular shotguns of all time. More than 10 million have been manufactured for hunting, self defence, law enforcement and also sporting uses.

ARMSCOR M30R6
COUNTRY OF ORIGIN
Philippines
DATE
1977
CALIBRE
18.5mm (12g)
WEIGHT
3.4kg (7.5lb)
OVERALL LENGTH
1016mm (40in)
FEED/MAGAZINE
Pump; 4-round magazine
RANGE
50m (54.68yds)

RIGHT: The legendary John Browning was responsible for the designs of some of the world's most famous firearms, including machine guns, rifles and shotguns. He is shown here with a machine gun prototype.

The Model 1300 Short Turkey is distinguished by its 457mm (18in) barrel.

The pump-action ARMSCOR M30R6 and M30 DI are manufactured by the Arms Corporation of the Philippines. The 12-gauge R6 is fed by a four plus one internal magazine, while the DI features a noticeably longer barrel at 660–711mm (26–28in) versus the 470mm (18.5in) of the R6.

Auto Shotguns

The world's first semiautomatic shotgun was the Browning Auto-5, introduced by the legendary John Browning in 1902 and produced until 1998. The name of the shotgun is derived from its semiautomatic action and the capacity of its tubular magazine. Manufactured in 12-, 16- and 20-gauge models, the Auto 5 is nicknamed the 'Humpback' due to the abrupt angle formed where the receiver and the stock are joined together.

Interestingly, the Auto 5 was discontinued in 1903, the same year of its appearance on the market, and then reinstated in 1923. Well-known for a powerful recoil, the shotgun has nevertheless been popular with hunters due to its reliability. During an era when cardboard shells might clog another shotgun, the Auto 5 remained in service. A grouping of friction rings is in place to prevent recoil that might damage the shotgun, and the rings are adjusted based on the type of load being fired.

Fabrique Nationale of Belgium produced the Auto 5 initially. During the 1940s, including weapons supplied during World War II, more than 850,000 were manufactured in the United States by Remington. The successor to the venerable Auto 5 is the Browning Gold series of gas-operated auto-loading shotguns. The Gold series includes target and hunting models in 10-, 12- and 20-gauges.

In the 1960s, the Remington Model 1100 addressed continuing issues with shotguns such as recoil, weight and durability in an unforgiving outdoor environment. The all-time best selling auto-loader in U.S. history, the Model 1100 incorporates a gas-operated mechanism that substantially reduces recoil and holds the record for consecutive shots fired without a

BENELLI AUTOLOADER M4

COUNTRY OF ORIGIN
Italy
DATE
1998
CALIBRE
18.5mm (12g)
WEIGHT
3.82kg (8.42lb)
OVERALL LENGTH
885mm (34.8in)
FEED/MAGAZINE
Gas operated; 7-round internal magazine
RANGE
50.2m (55yds)

failure. Fed by a four-round tubular magazine, it comes in 12-, 16-, 20- and 28-gauge, and .410 bore.

Italian firearms manufacturer Benelli was founded in 1967 and rapidly became recognized around the world as an outstanding producer of shotguns and other weapons. The Benelli Autoloader series encompasses a number of 12- and 20-gauge semiautomatic shotguns that are popular among hunters and law enforcement organizations. Benelli autoloaders are distinguished by their inertia operating system rather than relying on gas to eject a spent

RIGHT: German officers are shown with their shotguns following a successful effort to augment their food supply. The shotgun, with its short-range power, became popular in the trenches of the Western Front during World War I.

shell, eliminating the need to clean gas ports. Prominent in the Benelli line are the M3 and M4 Super 90.

Combat Shotguns

Pump action, semiautomatic and automatic shotguns have found military applications, particularly with the advent of trench warfare during World War I and the close quarter, sometimes house-to-house urban combat experienced in wars of the twentieth century.

Along with the famed Browning Auto 5 that performed admirably during the Great War and even prompted the German government to lodge a protest over the weapon based on conventions of the day, modern weapons such as the Mossberg 500 series and the Mk 1 variant of the Remington Model 870 have followed suit with combat shotgun variants. Other notable modern combat shotguns include the Atchisson Assault Rifle, also commonly referred to as the AA-12, the Franchi SPAS Models 12 and 15, the Beretta RS 200 Police Model, Benelli M4/M1014, the Daewoo USAS-12 auto shotgun, the Pancor Jackhammer and the Armsel Striker.

The original 12-gauge blowback Atchisson Assault Rifle was developed in the early 1970s by designer Maxwell Atchisson and gained notoriety with its reduced recoil, firing shells from a five-round box magazine. In 1987, Atchisson sold the rights to the shotgun to Military Police Systems, which has continued to improve the weapon with nearly 200 alterations. The current 12-gauge AA-12 is gas operated with a locked breech and fires up to 300 rounds a minute from an eight-round box magazine or 20- or 32-round drum magazines.

MOSSBERG 535
COUNTRY OF ORIGIN
United States
DATE
2005
CALIBRE
18.5mm (12g)
WEIGHT
3.06kg (6.75lb)
OVERALL LENGTH
1225.55mm (48.25in)
FEED/MAGAZINE
Pump; 5-round magazine
RANGE
40m (43.74yds)

ATCHISSON
COUNTRY OF ORIGIN
United States
DATE
1972
CALIBRE
18.5mm (12g)
WEIGHT
5.2kg (11.46lb)
OVERALL LENGTH
991mm (39.02in)
FEED/MAGAZINE
Blowback; 8-round box or
32-round drum magazine
RANGE
100m (109.36yds)

BERETTA RS 200
COUNTRY OF ORIGIN
Italy
DATE
1970
CALIBRE
18.5mm (12g)
WEIGHT
3.0kg (6.61lb)
OVERALL LENGTH
520mm (20.47in)
FEED/MAGAZINE
Pump; 6-round tube magazine
RANGE
100m (109.36yds)

BENELLI M4/1014
COUNTRY OF ORIGIN
Italy
DATE
1998
CALIBRE
18.5mm (12g)
WEIGHT
3.82kg (8.42lb)
OVERALL LENGTH
885mm (34.8in)
FEED/MAGAZINE
Gas operated; 7-round internal
tube magazine
RANGE
50.2m (55yds)

The Daewoo USAS-12 (Universal Sports Automatic Shotgun) is a derivative of the AA-12 that was manufactured in South Korea during the 1980s. The USAS-12 is a fully automatic 12-gauge combat shotgun that fires up to 450 rounds a minute from 10-round detachable box or 20-round drum magazines. A semiautomatic version has been offered on the civilian market; however, its distribution is severely restricted. A selective-fire variant is available to military and police organizations. Popular in Asia, more than 30,000 examples of the AA-12 were produced during the mid-1980s.

The short barrel pump-action Beretta RS200 is lightweight, distinguished by its wooden stock in the basic model and is capable of firing a variety of projectiles, including buckshot, standard slugs and rubber bullets for riot control. The military and security variant is designated the RS-202M1 and includes a folding stock, while the RS-202M2 includes a flash suppressor and heat shield. The military variants incorporate a feature in which the operator may load a single round into the chamber without operating the pump action.

Armsel Striker

Also known as the Protecta, Protecta Bulldog and Sentinel Arms Co. Striker 12 depending on the configuration, the South African Armsel Striker is the original of a series of rotating-cylinder combat shotguns designed by Hilton Walker in the 1980s and subsequently reengineered. A spring action rotated the 12-round cylinder in the early Striker. However, the slow operation of the firing mechanism was criticized, and this was replaced in 1989 with an automatic cartridge ejection system. This variant took the

name Protecta. A seven-round cylinder functions in the compact version. The Striker is a 12-gauge weapon that is currently used by police forces in Israel and by the South African National Defence Force.

The Benelli M4/1014 is also known as the M4 Super 90. Manufactured by the Italian firm of Benelli Armi SpA, the M4/1014 is a 12-gauge

BELOW: South African police officers patrol with loaded shotguns during a period of unrest. The shotgun has long been a standard weapon for riot control usage.

FRANCHI SPAS MODEL 12

COUNTRY OF ORIGIN
Italy
DATE
1979
CALIBRE
18.5mm (12g)
WEIGHT
4.4kg (8.75lb)
OVERALL LENGTH
1041mm (41in)
FEED/MAGAZINE
Pump; 8-round internal tube magazine
RANGE
40m (43.74yds)

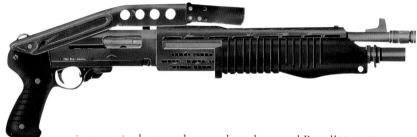

FRANCHI SPAS MODEL 15

COUNTRY OF ORIGIN
Italy
DATE
1996
CALIBRE
18.5 (12g)
WEIGHT
3.9kg (8.6lb)
OVERALL LENGTH
1000mm (39.37in)
FEED/MAGAZINE
Pump; 8-round detachable box magazine
RANGE
40m (43.74yds)

semiautomatic shotgun that employs the noted Benelli inertia driven or ARGO (auto regulating gas operating) system. In response to the requirements of the U.S. Army for a new 12-gauge combat shotgun, Benelli specifically developed the M4/1014, which prevailed over several competitors and entered service first with the U.S. Marine Corps when 20,000 examples were delivered in 1999. One versatile aspect of the shotgun is that it self-regulates for use with different types of ammunition.

An internal tube magazine feeds the M4/1014, five plus one for civilian models and seven plus one for the military version. The Picatinny sight rail accommodates both high-powered standard and night vision sights while leaving the basic ghost ring sights in place. To date, the M4/1014 has seen action with the U.S. military in Iraq and Afghanistan.

In 1982, another Italian arms manufacturer, Franchi SpA, introduced the SPAS-12, a 12-gauge pump-action, gas-actuated semiautomatic shotgun capable of firing a burst of four rounds in a single second. The SPAS-12 combat shotgun functions both in semiautomatic and pump-action modes with the latter most often used to fire less-than-lethal bean bag rounds or tear gas canisters. The gun is fed by an eight plus one internal tube magazine.

Design work was undertaken on the SPAS-12 in the early 1970s, and the shotgun has been sold to military and police organizations around the world. It is seen regularly in the media, particularly in feature films and video games. One inventive feature of the SPAS-12 is a magazine cutoff that interrupts the

semiautomatic action to allow the operator to insert and fire another type of round individually without being required to empty the magazine first.

Various chokes and a gas grenade launcher are fitted into the threaded end of the SPAS-12 barrel for deployment. Folding-stock variants are in circulation as well, and these feature a hook on the deployed stock that can be rotated to fit under the operator's forearm, allowing him to fire one handed, from cover or in other situations in which he might ordinarily be compromised.

The Franchi successor to the SPAS-12 is the SPAS-15 combat shotgun, which was produced for nearly 20 years beginning in 1986. The SPAS-15 is, like its predecessor, a 12-gauge pump-action, gas-actuated weapon. However, it is fed by detachable box magazines of three, six or eight rounds rather than the internal tube magazine of the SPAS-12. Its selective fire feature is engaged by depressing a button above the foregrip and sliding the foregrip slightly backward or forward.

The screw-in barrel attachments are retained in the SPAS-15, and the barrel is chrome-lined for better durability in combat conditions. The SPAS-15 is currently utilized by armed forces and police units in countries such as Belarus, Tunisia, Brazil, Portugal and Serbia. It was adopted by the Italian Army for service in 1999.

Pancor Jackhammer

One notable combat shotgun that showed early promise but ultimately failed to enter production was the Pancor Jackhammer, designed by American John Anderson in the mid-1980s. The 12-gauge, gas-operated automatic shotgun was capable of firing up to 240 rounds a minute from a 10-shell cylinder. Patented

PANCOR JACKHAMMER
COUNTRY OF ORIGIN
United States
DATE
1987
CALIBRE
18.5mm (12g)
WEIGHT
4.57kg (10.08lb)
OVERALL LENGTH
787mm (30.98in)
FEED/MAGAZINE
Gas operated; 10-round cylinder
RANGE
40m (43.74yds)

FEDERAL RIOT GUN
COUNTRY OF ORIGIN
United States
DATE
1933
CALIBRE
37mm (1.45in)
WEIGHT
2.5kg (5.51lb)
OVERALL LENGTH
730mm (28.74in)
FEED/MAGAZINE
Single shot
RANGE
40m (43.74yds)

in 1987, the Jackhammer was actually constructed on quite a limited basis with only a few prototypes available for testing. The U.S. military establishment rejected it and initial interest from other countries soon faded, forcing Anderson's Pancor company into bankruptcy.

While the Pancor Jackhammer operated on the time-tested revolver principle, many of its components were constructed of a rynite polymer material to vastly reduce the weapon's weight. Blending old and new technology, the shotgun fired with characteristics similar to early Webley-Fosbery and Mosin-Nagant pistols.

Riot Control Weapons

For many years, the term 'riot gun' was used as a general description for a shotgun of virtually any description. However, during the modern era the riot-control shotgun has evolved into a specific genre of the weapon that shares some similarities with its close cousins in the combat role.

The Federal Riot Gun (FRG) was developed in the United States in the 1930s by Pennsylvania-based Federal Laboratories. Manufactured as a single-shot break-action weapon, its ammunition was commonly the 37mm (1.45in) and 38mm (1.49in) baton or tear gas round. Familiar as a riot control weapon during the 1960s and 1970s, the FRG was seen on college campuses and in the streets of major American cities during periods of unrest. It also became the primary riot-control weapon of the British Army during the disturbances in Northern Ireland.

ARWEN 37
COUNTRY OF ORIGIN
United Kingdom
DATE
1977
CALIBRE:
37mm (1.45in)
WEIGHT
3.1kg
OVERALL LENGTH
710mm (28in)
FEED/MAGAZINE
Gas operated; 5-round rotary drum magazine
RANGE
40m (43.74yds)

Intended for non-lethal engagement with demonstrators, the FRG was eventually issued with modified firing instructions. Aiming and firing directly at an individual was determined to possibly result in lethal impact. Therefore, the FRG was often pointed at a paved road to ricochet into the target. Numerous police organizations around the world still carry the FRG in their arsenals.

In service with the British military, the FRG was supplanted by the ARWEN 37, a hybrid riot-control weapon that fires bean bag and tear gas rounds from a 37mm (1.45in) rotating drum, which prevailed over semiautomatic and pump-action prototypes in trials. The ARWEN (Anti Riot Weapon Enfield) 37 was designated to replace the FRG in 1977. Its drum holds five rounds, and the ARWEN 37 is currently manufactured under license in Canada. The ARWEN 37S 'Shorty' is a short-barrel compact version of the larger shotgun, while the ARWEN Ace is another single-shot 37mm riot control weapon.

The British Schermuly riot gun is a 38mm (1.49in) single-shot break-action smoothbore weapon that may also be fitted with an adapter barrel to fire conventional 12-gauge ammunition. Based on the design of a World War II-era flare pistol, it is constructed of aluminum components with a wooden stock by the firm of Webley and Scott. It offers the unique option among break-action riot weapons of being affixed to light armoured vehicles deployed in a security role.

Another familiar riot control weapon is the Smith & Wesson Model 276 gas gun. Chambered to accept the 37mm (1.45in) tear gas canister and other cartridges, it has been a familiar sight during civil unrest over the last half century. Its wooden stock and sling are recognizable attributes of the Model 276.

The semiautomatic Belgian FN 303 is a less-than-lethal riot weapon that fires its projectile with compressed air from a 15-round detachable drum magazine. Manufactured by Fabrique Nationale de Herstal, the FN 303 was designed by the Monterey Bay Corporation and built to be affixed to

SCHERMULY
COUNTRY OF ORIGIN
United Kingdom
DATE
1950
CALIBRE
38mm (1.49in)
WEIGHT
2.7kg (6lb)
OVERALL LENGTH
82.8mm (32.6in)
FEED/MAGAZINE
Single shot
RANGE
40m (43.74yds)

FN 303
COUNTRY OF ORIGIN
Belgium
DATE
2003
CALIBRE
18mm (.71in)
WEIGHT
2.3kg (5.07lb)
OVERALL LENGTH
740mm (29.1in)
FEED/MAGAZINE
Compressed air; 15-round
detachable drum magazine
RANGE
70m (77yds)

FABARM BETA
COUNTRY OF ORIGIN
Italy
DATE
1980
CALIBRE
18.5mm (12g)
WEIGHT
3.18kg (7lb)
OVERALL LENGTH
710mm (27.95in)
FEED/MAGAZINE
Side by side; double shot
RANGE
50m (54.68yds)

the M16 assault rifle when its stock is removed. Such a combination theoretically produces a less lethal alternative with a lethal weapon in the event that either should be needed in a given situation. The rotating barrel magazine also allows multiple types of riot-control ammunition to be loaded.

A serious incident in Boston in 2004 resulted in the death of a woman struck in the eye by a projectile fired from an FN 303. A lawsuit against the city and FN Herstal followed and settlement was reached. However, the city decided to destroy those FN 303s still in service due to their potential lethality. Despite this the FN 303 has been in production since 2003 and has been deployed in Iraq and Afghanistan as well as other global hot spots.

Double Barrel Sporting Shotguns

Skeet, trap or sporting clays have been popular competitive sports for many years, while the hunting of game, large and small, has served as a pastime and a means of sustenance as well. The shotgun, in its many and varied forms, has been central to each of these and is destined to remain so. Along with such classics as the Winchester Model 21, double barrel sporting shotgun designs abound.

Fabarm, located in Brescia, Italy, has manufactured the 12-gauge side-by-side break-action Beta with a checkered wooden stock, and it has been used for clay target shooting in the Olympic Games and other major competitions, while Fausti, also of Brescia, produces an outstanding line of both over/under and side-by-side shotguns for hunting and sporting pursuits. Produced by Mirocchi of Sarezzo, Italy, the 12-gauge over/under Model 99

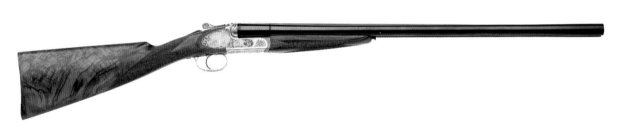

includes a hand-finished receiver. The 12-gauge over/under Miroku MK70 Sporter Grade 1 is versatile and effective with a variety of clay targets, employing interchangeable invector chokes and a satin varnished stock of American grade 2 wood.

The popular over/under shotgun configuration is favoured by hunters and sport shooting enthusiasts alike. Among the best-known over/under shotguns are a pair from Browning, the Superposed B25 and B125. In the autumn of 1923, John Browning received the first of two patents on the superposed over/under shotgun design. According to Browning corporate, it was his last invention. Eight years later, the configuration was introduced into the company's commercial line.

The break-action B25 was an early over/under Browning shotgun produced in Belgium by Fabrique Nationale. Introduced in 1926, the B25 was originally made only for 12-gauge 69.8mm (2.75in) shells. In 1940 it was offered in 20- and 28-gauge and .410 bore. Grades vary from basic to the highly embellished Midas. The rifle is constructed with a box lock and single trigger, and the safety catch located along the neck also works as a barrel selector. The Browning B26 Liege entered production in the early 1970s as a less expensive option to the B25, and the semiautomatic B80 was marketed from 1980 to 1988 with components manufactured by Beretta of Italy.

In 1984, Browning introduced the B125 ostensibly as a replacement for the B25; however, a public outcry arose and the two shotguns were marketed at the same time. The Japanese firm Miroku manufactured parts for the B125, and the final assembly was done by the Browning custom shop in Belgium. The shotgun is known for its precise tolerances, and the stock is of exquisite French walnut. Hunting models include the Special Chasse, while the Parcour de Chasse is designed for sporting clays with a heavier barrel and wider barrel ribs.

Toward the end of its general production run additional B125 models were produced, including the Special Chasse Europe, the 20-gauge Superlight and the Trap F1. The 325, 425 and 525 over/under trap shotguns are still

BROWNING B125
COUNTRY OF ORIGIN
Belgium
DATE
1984
CALIBRE
18.5mm (12g)
WEIGHT
3.46kg (7.625lb)
OVERALL LENGTH
1085.85mm (42.75in)
FEED/MAGAZINE
Over/under; double shot
RANGE
50m (54.68yds)

considered ideal for the sport. By 2001, however, Browning had dropped the entire line.

One of the most influential over/under shotgun families of the twentieth century is the Beretta SO series, also known as the Sovrapposto. After World War I, the over/under began to take centre stage among its peers. British manufacturers such as Boss and Woodward, and the Browning Superposed being manufactured in Belgium, ushered in a wave of renewed interest in the over/under. At Beretta, the idea for an elegant entry in the over/under market emerged, and it would include a locking system seen previously on side-by-side guns along with geometric barrel shoulders that locked up firmly with the barrel walls for a sturdy gun. Beretta's signature monobloc barrels were also a feature.

Beretta SO1

Despite its innovation, the SO did not take complete shape until the early 1930s and it was not introduced to the market until 1935. Intended for both competitive shooting and hunting, the SO series first appeared in 12-gauge with an automatic ejector and double trigger. After the interruption of World War II, production resumed and the original shotgun was designated the SO1. The SO2 Super Caccia and the SO3 competition model followed. By the 1960s, the SO4 had joined the fold. The SO4 Trap model was introduced in 1971, and in the late 1980s the SO5 was a fixture at skeet, trap and sporting clay events. The SO6 has garnered numerous awards and is well known in its sporting variant.

The series entered a new era in 1990, when the SO9 became the first of its family to be available in small gauge, including 20- and 28-gauge, and .410 bore. Although the SO9 maintains the look and feel of the larger members of the series, it opened the experience of ownership up to more enthusiasts with a lower price point.

In 2004, the S010 was introduced with a new design that dispensed with the traditional Beretta cross bolt locking system in favour of a combination

BERETTA SO9
COUNTRY OF ORIGIN
Italy
DATE
1990
CALIBRE
15.6mm (20g)
WEIGHT
2.44kg (5.375lb)
OVERALL LENGTH
1160mm (45.7in)
FEED/MAGAZINE
Over/under; double shot
RANGE
50m (54.68yds)

Beretta system that is influenced by other manufacturers' handiwork. The SO10 includes a titanium trigger blade and detachable locks removed by taking off the heads of concealed pins.

The Beretta competition grade 682 over/under shotgun is manufactured in 12-, 20- and 28-gauges, as well as .410 bore. In production since 1984, several grades for sporting clays, trap and skeet are marketed. Although they share numerous common parts, two primary versions of the 682 have been manufactured. The original was made through 1994 and is known as the wide or large frame 682.

At some point – probably to save weight in competition guns – Beretta reduced the barrel weight, added a new choke system and narrowed the receiver to complete the more recent version. Those who attempt to combine

BERETTA 682
COUNTRY OF ORIGIN
Italy
DATE
1984
CALIBRE
18.5mm (12g)
WEIGHT
3.4kg (7.5lb)
OVERALL LENGTH
1160mm (45.7lb)
FEED/MAGAZINE
Over/under; double shot
RANGE
50m (54.68yds)

BERETTA ULTRALIGHT
COUNTRY OF ORIGIN
Italy
DATE
1990
CALIBRE
18.5mm (12g)
WEIGHT
2.6kg (5.8lb)
OVERALL LENGTH
1070mm (42.5in)
FEED/MAGAZINE
Over/under; double shot
RANGE
50m (54.68yds)

SALVINELLI L1
COUNTRY OF ORIGIN
Italy
DATE
2012
CALIBRE
18.5mm (12g)
WEIGHT
3.74kg (8.25lb)
OVERALL LENGTH
1178mm (46.375in)
FEED/MAGAZINE
Over/under; double shot
RANGE
50m (54.68yds)

parts from an original pattern weapon with a newer 682 should proceed with caution as ejectors and other components may not perform properly.

The Beretta Ultralight introduced a lighter and shorter over/under shotgun to the market. The 12-gauge Ultralight stock is 368mm (14.5in) long, a full 762mm (30in) shorter than the average for all shotguns. Its barrel length is 711mm (28in), and the Ultralight is also approximately 454g (1lb) lighter than the average shotgun.

Perazzi MX Series
The high-end Perazzi MX series includes the over/under 12-gauge MX8 with an exhibition grade French walnut stock and gold inlay. A dropout trigger group, ported barrels and interchangeable choke tubes make it ideal for sporting clays or high level skeet competition. The break-action 12-gauge Lanber Deluxe Sporter is available in sporting, hunting and game variants. Lanber, a division of Winchester Australia, offers the Deluxe Sporter with chrome-lined barrel, wide rib and flush fitting multichokes as standard. Laser checkering is featured on the wooden stock and foregrip.

The Armi Salvinelli L1 over/under break-action 12-gauge shotgun utilizes a Boss-style locking system that is also similar to that of Perazzi. The three Salvinelli brothers entered the firearms manufacturing business in 1955 and became known for producing high quality competition shotguns. Interchangeable chokes and a single selective trigger are options with the L1, and the wooden stock is custom-made to the buyer's specifications. Depending on the grade, the receiver finish is glossy blued, dull chrome plated or case hardened. The forend is either in beavertail or round style.

Thanks to a simplified, less expensive manufacturing process, the popular Ruger Red Label over/under shotgun has recently returned to the market with reduced recoil compared to the original model, along with a redefined centre of gravity while retaining the walnut stock of the earlier production run. Constructed for hunting, sporting clays or skeet, the Red Label was introduced in 1977 and withdrawn from the Ruger line in 2011. Initially available in 12-gauge only, the 20-gauge model was set to return to the market in 2014.

ABOVE: The Ruger Red Label over/under shotgun reached the market in 1977 and was briefly withdrawn in 2011. With its return, the Red Label is available in both 12-gauge and 20-gauge.

BAIKAL IZH-27/ REMINGTON SPR-310

COUNTRY OF ORIGIN
Russia
DATE
1977
CALIBRE
18.5mm (12g)
WEIGHT
3.49kg (7.7lb)
OVERALL LENGTH
1244.6mm (49in)
FEED/MAGAZINE
Over/under; double shot
RANGE
50m (54.68yds)

RIGHT: Krieghoff
International is now in
its fifth generation of
gunmakers, spanning 125
years of excellence. The
company was founded in
Ulm, Germany, in 1886
and is known for precision
shotguns such as this
over/under.

The Model 90 over/under was manufactured by the Marlin Firearms Company of New Haven, Connecticut, from 1937 to 1963, and one production run was completed for Sears Roebuck as an affordable, no frills and straightforward hardware gun. Sears Roebuck marketed the Model 90 under the name Ranger until 1941 and under the J.C. Higgins moniker from 1946 until it was discontinued. The Model 90 was made available in 20-gauge and .410-bore versions.

In 1977, Remington introduced the Spartan 310 over/under shotgun. The standard SPR-310, as it is known, is manufactured in a nickel-plated or blued receiver with barrels of 660mm or 711mm (26in or 28in), while the sporting model SPR-310S is distinguished by its 749mm (29.5in) barrel. Imported from Russia by Remington from 2005 to 2009, the break-action SPR-310 is a close relative of the IZh-27 shotgun manufactured by the Izhevsk Mechanical Plant and referred to as the Baikal. It was produced in 20-gauge and .410 bore, as well as the better-known 12-gauge.

The Weatherby Athena over/under shotgun is chambered for 76.2mm (3in) 12- and 20-gauge and .410 bore magnum shells, while the 28-gauge is chambered for only 69.8mm (2.75in) shells. The .28-gauge and .410 are constructed with integrated chokes, and the 12- and 20-gauge models are completed with interchangeable chokes. The straight grip stock and forend are typically checkered walnut, and double triggers are common.

Fabbrica Armi Isidoro Rizzini

A number of shotguns from Italian firearms manufacturer, FAIR (Fabbrica Armi Isidoro Rizzini) of Brescia, Italy, are prominent on the over/under market. The Premier is purpose built as a basic shotgun for beginners. Utilizing the same locking block as other FAIR shotguns, it is available in 12-, 20- and 28-gauge, and .410 bore. The model is produced in interchangeable or fixed chokes and is built with a single selective trigger or double trigger. Wood furnishings are of standard walnut, and the breechblock is blued. The Premier EM is equipped with an automatic shell ejector, while

the Premier Deluxe features simple engraving on the breechblock.

The FAIR Jubilee family of shotguns includes several models that have been in production for some time that have been renamed in recent years. For example, the FAIR 700 includes variants that are still also recognized by their earlier names, such as the LX-680 and SRC-620. The Jubilee 700 series includes barrel block locking and chambers for the 69.8mm (2.75in) and 76.2mm (3in) 12-gauge ammunition. The breechblock of the companion 702 is chromed and features simple engraving, while the box lock is blued, finished in matte black and lined in gold. Interchangeable or fixed chokes are available. The Jubilee 702 includes trap, skeet and sporting clay models with barrel lengths appropriate for each.

The FAIR Jubilee 900 is an upscale version of the 702, which was formerly known as the LX-900 Caccia. In turn, the Jubilee 902 is an embellished version of the 900 with finely engraved side plates on its box lock. Prestige models in these FAIR shotguns are the top-of-the line offerings from the company. The Prestige 702 includes a chromed breechblock and side plates

FAIR JUBILEE PRESTIGE
COUNTRY OF ORIGIN
Italy
DATE
1971
CALIBRE
18.5mm (12g)
WEIGHT
3.2kg (7.05lb)
OVERALL LENGTH
1066.8mm (42in)
FEED/MAGAZINE
Over/under; double shot
RANGE
50m (54.68yds)

BROWNING GOLD HUNTER
COUNTRY OF ORIGIN
United States
DATE
1995
CALIBRE
18.5mm (12g)
WEIGHT
3.4kg (7.5lb)
OVERALL LENGTH
1232mm (48.5in)
FEED/MAGAZINE
Gas operated; 4-round tubular magazine
RANGE
50m (54.68yds)

FABARM LION H38 HUNTER
COUNTRY OF ORIGIN
Italy
DATE
2000
CALIBRE
18.5mm (12g)
WEIGHT
3.05kg (6.762lb)
OVERALL LENGTH
1219.2mm (48in)
FEED/MAGAZINE
Gas operated; 8-round tubular magazine
RANGE
50m (54.68yds)

and the Prestige Tartaruga includes a tempered breech and side plates, fine engraving and gold inlay.

Some over/under shotguns fall into the category of combination guns. With at least two barrels, the combination gun includes one shotgun barrel and a second rifle barrel. These are most often utilized as hunting guns and allow the hunter to engage a variety of game, large and small. One fine example of the combination gun is manufactured by Pennsylvania-based Krieghoff International, a division of Heinrich Kreighoff of Ulm, Germany. Featuring a 16-gauge shotgun barrel joined with a 7mm (.278in) rifle barrel, the Krieghoff is fitted with a cheek pad and often matched with a Zeiss 4x telescopic sight affixed by claw to the right-hand side for removal.

Semiautomatic Sporting Shotguns

An array of gas-operated semiautomatic shotguns is available on the market today, and among them are numerous models from some of the world's most prominent firearms makers. The Browning Gold Hunter is one of several variants to the Gold Shotgun introduced in 1995 and specifically oriented for either championship target shooting or hunting. The Gold Hunter in 12-gauge utilizes a 660mm (26in) barrel with vent ribs and three optional screw-in choke tubes. It has a magazine capacity of either four plus one in the chamber or three plus one in the chamber depending on the size of the shell.

The Gold Hunter's wood stock is finely grained walnut with checkering on the pistol grip and forend. It has a reputation for easy cleaning and quick disassembly of the self-adjusting gas system. When the bolt is locked rearward, as when the magazine and chamber are empty, the speed load

feature allows a shell inserted into the magazine tube to move immediately to the chamber, readying the shotgun to fire immediately.

Two semiautomatic shotguns from Fabarm, the popular 12-gauge H35 Azur with triwood stock and gas operation, and the Lion H38 Hunter, are notable. The H38 Hunter features a tribore barrel with inner choke and an eight-shell magazine where allowed by law. Three optional screw-in chokes are available, and the receiver is finished in black while the wooden stock is polished to a high sheen.

Beretta has produced its AL391 semiautomatic shotgun since 1990 as the successor to the AL390. The AL391 has developed a reputation for excellence in hunting birds and shooting skeet or trap. Although it is mechanically similar to its predecessor, the AL391 has a slightly different shaped stock, an aluminum receiver reduces overall weight and it is slimmer toward the forend.

The AL391's convenient magazine cutoff allows a new shell to be chambered individually without emptying the magazine first, while the magazine holds three rounds plus one in the chamber and an optional magazine plug is available to restrict capacity in compliance with prevailing hunting laws. The self-compensating gas operated action of the AL391 is more complicated than that of other shotguns; however, it does automatically adjust for shells of different charges and different strength of recoil.

Available in 12- or 20-gauge, the AL391 is sold in four models, the standard Urika, the upscale Teknys with finer finishes and engraving, and the Xtrema, specifically designed for hunting waterfowl and produced in 12-gauge only with the ability to accept heavier magnum shells. The heavier Xtrema 2 was introduced in 2004 in response to the demand for a versatile all-around shotgun.

The Xtrema 2 is stripped and cleaned more easily due to a reduction in springs and O-rings. Both fewer wearing parts and a longer barrel tang, to improve the security of the barrel to the receiver, enhance accuracy, and the barrel is overbored to reduce shot climb and improve its pattern. Old-

BERETTA AL391 URIKA

COUNTRY OF ORIGIN
Italy
DATE
1990
CALIBRE
18.5mm (12g)
WEIGHT
3.31kg (7.3lb)
OVERALL LENGTH
1295.4mm (51in)
FEED/MAGAZINE
Gas operated; 3-round magazine
RANGE
50m (54.68yds)

BENELLI SUPER BLACK EAGLE II

COUNTRY OF ORIGIN
Italy
DATE
2009
CALIBRE
18.5mm (12g)
WEIGHT
3.31kg (7.3lb)
OVERALL LENGTH
1260mm (49.6in)
FEED/MAGAZINE
Inertia driven; 3-round magazine
RANGE
50m (54.68yds)

FRANCHI I-12

COUNTRY OF ORIGIN
Italy
DATE
2008
CALIBRE
18.5mm (12g)
WEIGHT
3.49kg (7.7lb)
OVERALL LENGTH
1194mm (47in)
FEED/MAGAZINE
Inertia action; 3-round magazine
RANGE
50m (54.68yds)

time hydraulic shock absorption reduces the recoil felt by the operator by a whopping 44 per cent. The 12-gauge Xtrema 2 is fed by a tube magazine of up to 11 rounds that is located under the barrel. Metal parts that aren't aluminum or stainless steel are protected from corrosion by a sealing membrane.

The Benelli Super Black Eagle II inertia-driven semiautomatic shotgun features an ergonomic trigger guard and safety along with a removable trigger assembly. Its barrel and choke tubes are cryogenically treated for a more uniform shot pattern, and proprietary Airtouch checkering is molded into the synthetic stock for a firm grip. The ComforTech recoil reduction system reduces standard recoil from the 12-gauge variant by 48 per cent. In addition to a variety of shells, the Super Black Eagle II fires a 76.2mm (3in) slug.

The Benelli Vinci SuperSport semiautomatic shotgun is described as a 'race gun' that is well appointed for any clay competition. The 12-gauge SuperSport utilizes a cryogenically treated barrel and chokes for improved accuracy and features a sleek carbon fibre stock with blued and bright finishes on metal components. The barrel is fixed, and the recoil involves the bolt unlocking and cycling the action. Therefore, the SuperSport provides such a rapid firing cycle that it is well known for its speed and results in a lower recoil into the shoulder of the operator than perhaps any other 12-gauge shotgun.

The inertia action semiautomatic Franchi I-12 is the result of collaboration between Franchi and fellow Italian firearms manufacturer Benelli, and the ejection system requires no adjustment and operates cleanly. The 12-gauge shotgun is fed by a four-round magazine, and its stock is in satin-finished walnut, or solid black or camouflage synthetic. Laser checking assists with a firm grip and five screw-in choke tubes are available.

Bolt-action shotguns are rare. However, Mossberg has made several over the years, including the .410-bore Model G Repeater that was produced beginning in 1933. A four-shot gun used primarily to hunt rabbits and squirrels, the Model G was followed by the Mossberg Model 70, a single-shot bolt-action shotgun that could be broken down into two pieces for transport. In 1939, the Model 70 was produced in a 20-gauge version, and by the beginning of World War II the company was selling at least a dozen different bolt-action shotguns. During the 1950s the demand for Mossberg bolt-action rifles revived, and the .410-bore Model 183 and 20-gauge Model 185 sold briskly. The 16-gauge Model 190 and 12-gauge Model 195 soon followed.

The Model 695 was the last in the Mossberg bolt-action line and was introduced in 1995. Before the bolt-action line was discontinued in 2003, the Model 695 was recognized as a niche gun, relatively expensive to produce and appealing to a narrow group of enthusiasts. One version was a slug-firing shotgun with a rifled barrel and adjustable fibre-optic sights.

ABOVE: The Benelli SuperSport semiautomatic shotgun has a fine reputation in sporting clays. It has been described as a 'race gun' that provides a rapid firing cycle and low recoil.

The Tar Hunt RSG-12 single-shot bolt-action shotgun eclipsed the performance of other slug-firing shotguns, increasing range to more than 91m (100yds). It is reputed to handle and shoot like a rifle. Available in 12-, 16- and 20-gauge variants it is produced in high gloss metal, nickel Teflon and dipped camouflage stock finishes.

'The Knick'

Among single barrel shotguns, the Model 4E trap gun was produced from 1926 to 1991 by the Ithaca Gun Company of Ithaca, New York. Designer Emil Flues of Bay City, Michigan, designed the lock utilized on Ithaca single-barrel models from 1914 to 1922. Subsequently, Frank Knickerbocker of the Ithaca company designed a line of single-barrel shotguns that were produced from 1922 through to 1988. These shotguns were known to many simply as 'The Knick.'

The single-barrel Linder Sextuple Trap shotgun received its name due to its six locking lugs. The 12-gauge box lock shotgun originated with Lindner in Suhl of Thuringia in Germany for Charles Daly of New York. It featured automatic ejectors and full choke barrels with full length ventilated ribs.

The Winchester series of single-barrel shotguns, including the Model 20, Model 36 and Model 37, were introduced following World War I and initially rushed into production to utilize excess capacity due to falling arms orders related to the cessation of hostilities. Made only in .410-bore and with a walnut stock, the Model 20 was quite a basic design. The breechloading bolt-action Model 36 was actually cocked with a pull of the head of the firing pin, and about 20,000 were sold during a production run from 1920 to 1927. Winchester marketed the Model 36 as a garden gun, to be used for pest control around homes and farms.

The Model 37 was produced from 1936 to 1963, and over a million of the basic, reasonably priced shotgun were purchased. It was manufactured in two variants, the standard and the boy's model, which was offered beginning in 1958. The break-action Model 37 was manufactured in 12-, 16-, 20- and 28-gauge, and .410-bore. During World War II a large number of U.S. National Guard troops were issued the Model 37.

One of the best-known lever-action shotguns is the Winchester Model 1887. Approximately 65,000 were manufactured after Winchester purchased the patent from John Browning and placed it in production. When the run ended in 1901, the shotgun had been manufactured in 10- and 12-gauge standard versions and in a riot gun variant.

RIGHT: The Winchester Model 1887 was a 10- or 12-gauge lever action shotgun fed from a 5-round tubular magazine. The Model 1887 was designed by John Browning at the request of the firearms manufacturer.

Sporting Rifles

The rifle holds a place unparalleled in modern history, shaping the course of events for the last 400 years like nothing else. Sporting rifles encompass the largest collective and descriptive term for these indispensable tools. Whether displaying keen marksmanship in target shooting or hunting for game, large and small, across fields and forests, sporting rifles enrich the human experience, allowing our species to flourish.

Many of the best-known sporting rifles in the world trace their origins to the genius of such men as the Mauser brothers, John Browning and others. Their ideas and practical application of them reach across the years. Today, with centuries of innovation on which to build, the development of new sporting rifles continues unabated as perfection is and always will remain elusive.

A veritable treasure-trove of outstanding sporting rifles is now available, and undoubtedly it is an impossible task to mention each in detail. Therefore, the broad view presents a variety of these and invites the reader to consider his or her own opinion as to strengths and weaknesses in the

LEFT: A hunter relaxes with his trusty sporting rifle resting on his shoulder. This particular rifle is embellished with fine engraving.

MAUSER 98
COUNTRY OF ORIGIN
Germany
DATE
1898
CALIBRE
7.62mm (.30in)
WEIGHT
4.09kg (9lb)
OVERALL LENGTH
1250mm (49.2in)
FEED/MAGAZINE
Bolt action; 5-round box magazine
RANGE
550m (601.49yds)

process. The sporting rifle captivates its owner and maintains its status – past, present and future – as an object of admiration, envy and romance.

Bolt-action Sporting Rifles

By the time the Mauser brothers had completed their iconic Model 1898 design, it was the culmination of five previous efforts, each building on its predecessor. The result was, quite arguably, the most influential bolt-action configuration of all time. The Mauser 98 not only served as the standard issue rifle of the German Army from 1898 to 1935, but has also risen to the highest echelon among hunting rifles.

The influence of the Mauser design is readily apparent in the simple fact that it has been interpreted, improved or modified countless times. During the twentieth century, virtually every bolt-action rifle of consequence bore the unmistakable imprint of Mauser influence. In truth, many of the so-called improvements may be traced to an actual effort to reduce production costs rather than an effort to better the smooth action of the rifle.

BELOW: A favourite among deer hunters, the Mauser 98 has proven its worth as both a military and sporting rifle chambered for a variety of rounds.

The most prominent attributes of the Mauser 98 include a machined one-piece bolt, and this consists of the handle, a pair of front locking lugs and a single auxiliary locking lug in the rear, a bolt mounted three position wing safety and a controlled feed extractor. Other significant attributes include a receiver-mounted ejector, a one-piece trigger guard and magazine assembly, a flat-bottomed receiver with an integral recoil lug and an internal staggered load cartridge box magazine. Many of the original 7.92mm (.312in) Mauser 98s that are used by hunters today have been rechambered for 8mm (.314in), 9mm (.35in) or 7.62x63mm (.30in) rounds.

New rifles patterned after the original Mauser 98 remain in development and production today. The overall quality of the precision-crafted rifle has allowed many vintage rifles to remain in the field, and the fixed ejector and controlled round feeding are ideal for hunting large game.

Winchester began the manufacture of its famed Model 70 in 1936, and the production run of the original rifle lasted until 1963. Dubbed the 'Rifleman's Rifle', the Winchester Model 70 began as a replacement for the earlier Model 54 that had been available in six variants. In 1964, Winchester launched a new Model 70, and its reception by the gun-owning public was less than enthusiastic. Particularly, the new Model 70 did not have the visual appeal of the original and its receiver incorporated less expensive stamped components. Further, the Mauser-style extractor was replaced.

For many, the original Model 70 was irreplaceable. It combined the very best of both the Mauser 98 and Model 54 to provide a sturdy and handsome rifle that sold in the millions. Produced in 18 calibres, the bolt-action magazine-fed Model 70 shares the turn bolt locking mechanism with the Mauser as well as the dual front locking lugs and full length extractor and fixed ejector mounted on the receiver. The open top receiver of both is machined with an integral recoil lug. The cone shaped breech of the Model 54 was retained for its ability to guide rounds into the firing chamber even with a slight misalignment.

For the operator, the Model 70 improved the Mauser's substantial firing pin fall with a much more rapid action and smoother report. The trigger

WINCHESTER 70
COUNTRY OF ORIGIN
United States
DATE
1936
CALIBRE
7.62mm (.30in)
WEIGHT
3.86kg (8.5lb)
OVERALL LENGTH
1162mm (45.75in)
FEED/MAGAZINE
Bolt action; 4-round internal magazine
RANGE
457.2m (500yds)

REMINGTON XR-RANGEMASTER

COUNTRY OF ORIGIN
United States
DATE
2005
CALIBRE
Various
WEIGHT
3.18kg (7lb) average
LENGTH (BARREL)
660mm (26in)
FEED/MAGAZINE
Bolt action; single shot
RANGE
300m (328yds)

REMINGTON 700 BDL

COUNTRY OF ORIGIN
United States
DATE
1962
CALIBRE
7.62mm (.30in)
WEIGHT
4.08kg (8.09lb)
OVERALL LENGTH
1055mm (41.5in)
FEED/MAGAZINE
Bolt action; 5-round internal magazine
RANGE
457.2m (500yds)

configuration has earned praise over the years as being one of the finest ever introduced. Among the best-known incarnations of the original Model 70 are the Featherweight, Super Grade and Standard. In 1992, Winchester decided to reintroduce the original rifle under the new name Model 70 Classic. It reached the market in 1994.

Among the vast number of outstanding bolt-action Winchester sporting rifles manufactured during the twentieth century are the Model 52, a target rifle that was in production from 1920 to 1980 and marketed in eight variants, including the Model 52 Sporter produced from 1934 to 1958, and the Model 72, a single shot bolt-action 5.58mm- (.22in-) calibre rifle marketed from 1938 to 1959 with a tubular magazine positioned under the barrel. More than 160,000 were produced.

The Model 67 and Model 69 were bolt-action sporting rifles introduced in the mid-1930s in 5.58mm- (.22in-) calibre. The Model 67 was made in four variants including a smoothbore. The Model 69 was devised as a combination target and hunting rifle at a modest price.

Remington Rifles

Remington has produced numerous bolt-action rifles through the years, and the most recognized of these is the 700 series, an enhancement of the Model 721, 722 and 725 rifles of the late 1940s. The Model 700 is the best selling sporting rifle of all time, and more than five million have been manufactured to date. Produced in more than a dozen variants, the Model 700 was designed by Mike Walker and Wayne Leek, who went on

to manage the Remington research and development department. The
Model 700 entered production in 1962, and the capacity of its internal
magazine was eventually up to six rounds depending on the calibre.

In 1969, a redesign effort resulted in an extended rear bolt shroud and
ushered in an era during which the Model 700 was rechambered for a variety
of ammunition, its stock was reshaped on several occasions and sights and
barrel lengths were modified many times. The Model 700 served as the base
rifle for the U.S. Marine Corps M40 sniper rifle that was deployed in large
numbers during the Vietnam War, and police variants have been produced
with Kevlar stocks rather than the highly polished walnut that has proven
so popular with all variants, including the Mountain, the common BDL
(B Deluxe Grade) and the CDL (C Deluxe Grade).

The Model 700 is considered by some to be the rifle that saved Remington.
The Model 721 was highly accurate but unattractive, and the Model 700

ABOVE: The Remington
Model 700 served as the
base weapon for the M40
sniper rifle used extensively
during the Vietnam War
by the U.S. Marines. This
photo depicts a sniper and
instructor during exercises.

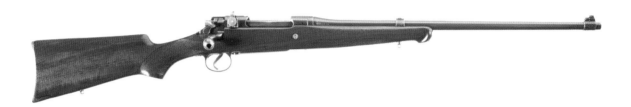

REMINGTON MODEL 30
COUNTRY OF ORIGIN
United States
DATE
1921
CALIBRE
7.62mm (.30in)
WEIGHT
4.17kg (9.3lb)
OVERALL LENGTH
1175mm (46.26in)
FEED/MAGAZINE
Bolt action; 5-round internal box magazine
RANGE
457.2m (500yds)

REMINGTON 510
COUNTRY OF ORIGIN
United States
DATE
1939
CALIBRE
5.58mm (.22in)
WEIGHT
2.18kg (4.8lb)
OVERALL LENGTH
1086mm (42.75in)
FEED/MAGAZINE
Bolt action; single shot
RANGE
91.44m (100yds)

was developed to retain the accuracy but improve the aesthetics and other aspects of the rifle. Remington found the right combination of reliability, functionality and beauty in the Model 700 while minimizing production costs in the process – a substantial feat indeed. Its construction with the famed 'three rings of steel' surrounding the cartridge head adds strength. The push feed system utilizes a circlip extractor and plunger ejector mounted in a recessed bolt face.

The fine trigger, rapid lock and round, open top receiver machined from stock steel are hallmarks of the Model 700. The superb accuracy of the Model 700 has also served for years as a benchmark for other rifles – those that are versatile enough to be confidently operated in the field or as a target rifle.

Remington had sought to capitalize on the abundance of factory workers and plenty of surplus rifle components when the contract production of the Model 1917 Enfield ended abruptly after World War I. The result was the bolt action 7.62x63mm (.30in) Model 30, a retooled version of the Enfield expressly for the civilian hunting market. It was introduced in 1922. The Model 30 proved too heavy and expensive to be commercially viable, and four years later the improved Model 30 Express, shorter and lighter than the original, became available.

The Remington bolt-action Model 34 target rifle reached the market in 1932, and nearly 170,000 were sold through 1936. The tubular magazine of the Model 34 was placed under the barrel and held up to 22 short 5.58mm- (.22in)- calibre rounds. The rifle was known for its light weight, and further variants such as the Model 34 NRA were equipped with several different types of sights.

REMINGTON 513T
COUNTRY OF ORIGIN
United States
DATE
1940
CALIBRE
5.58mm (.22in)
WEIGHT
4.08kg (9lb)
OVERALL LENGTH
1149mm (45.25in)
FEED/MAGAZINE
Bolt action; 10-round
detachable box magazine
RANGE
91.44m (100yds)

Among the early Remington bolt-action target rifles, those of the 500 series stand out. The single shot Model 510 Targetmaster reached the market in the spring of 1939, while nearly 125,000 examples of the Model 513T Matchmaster were sold in three decades from 1939 to 1968. The Model 513T featured highly accurate sights, a 10-round detachable box magazine, a 'floating' barrel, adjustable sling swivels for easy carry and a redesigned stock. Both the single shot Model 510 and the 513T were chambered for a variety of 5.58mm- (.22in-) calibre rounds, including .22in short, .225in, .221in, .22in long, .22in long rifle and .22in Remington cartridges.

The Remington Model 514A was introduced in 1948 as a lightweight .22in-calibre single shot addition to the 500 series, and by 1961 the Model 514BC, or boy's carbine, was introduced with a shorter stock and barrel.

In 1967, the Remington Model 788 entered the market. Designed by Wayne Leek, the rifle was chambered for numerous types of ammunition including 5.66mm (.223in), 6mm (.23in), 7.82mm (.308in), 7.62mm (.30in) and 11.1mm (.44in). Production ended in 1987 with 565,000 sold. Other notable Remington bolt action rifles include the Model 725, produced as a direct competitor with the rival Winchester Model 70 with a four-round internal magazine chambered for 7.62x63mm (.30in), 6.85mm (.27in) and 7mm (.28in) ammunition, the Model 600 push feed in standard and carbine variants produced from 1964 to 1967 and revived as the Model 660, Model Mohawk 600 and Model 673, the Model 541-S and the Model 7600.

Other Popular Bolt-action Rifles

A contemporary of the Mauser 98, the Mannlicher Schoenauer bolt-action rifle originated in the Kingdom of Austria-Hungary at the turn of the twentieth century. Issued to the Greek Army in 1903, the Mannlicher Schoenauer also served as the primary rifle of the Austro-Hungarian Army

MANNLICHER SCHOENAUER
COUNTRY OF ORIGIN
Austria-Hungary
DATE
1903
CALIBRE
6.5mm (.256in)
WEIGHT
3.83kg (8.44lb)
OVERALL LENGTH
1226mm (48.25in)
FEED/MAGAZINE
Bolt action; 5-round internal box magazine
RANGE
600m (660yds)

during World War I. In 1903, a sporting variant of the rifle was introduced to the civilian market concurrently with the military rifle that was expected to generate considerable export sales.

The Mannlicher Schoenauer was developed through a collaboration between Ferdinand Ritter von Mannlicher, who envisioned the basic rifle, and a rotary drum five-shot magazine that was conceived by Otto Schoenauer. Chambered for a 6.5mm (.256in) cartridge, the rifle grew in popularity in Europe and the United States prior to the Great War. A bolt with a pair of opposed front locking lugs with the handle serving as a safety lug contributed to the precision action of the Mannlicher Schoenauer, and the rifle featured a controlled cartridge feeding design along with an ejector that was fixed and mounted to the receiver.

The drum magazine of the Mannlicher Schoenauer dispenses with the drag associated with the magazine follower of other rifles, and the split rear receiver ring channels the bolt handle as it is withdrawn. Although the rifle's lock time is slow in comparison to others, and mounting a telescopic sight may be a challenge, the rifle is a smooth operator. Its magazine is loaded by straight stripper clips.

The Mannlicher Schoenauer rifle remained in production until 1971, and approximately 310,000 were manufactured. The sporting variants produced include carbine versions along with a conventional half-stock rifle. The rifle is often distinguished by its characteristic butter-knife bolt handle, and with its double set trigger the operator sets the front trigger with the rear trigger, facilitating a light release. Independent gunsmiths still build the Mannlicher Schoenauer; however, its high cost was a factor in the eventual termination of mass production.

Weatherby Mark V

In the mid-1950s, American Roy Weatherby began designing a series of rifle cartridges, including 5.68mm (.224in), 9.6mm (.378in) and 11.68mm (.46in) magnum, and then set about configuring a rifle that would

accommodate them. By 1957, his Mark V action was in the prototype stage and featured a comparatively huge bolt with nine lugs rather than the two that most rifles of the day included. The lugs themselves are smaller than those of other rifles, reducing bolt lift to only 45 degrees. The action of the bolt entering the receiver was reminiscent of a piston, and the fully enclosed case head made for an extra tight lockup.

Although the Weatherby Mark V shared an open top, flat bottom receiver with an integral recoil lug and the staggered-cartridge box magazine of the Mauser 98, the bolt configuration, devised by Weatherby engineer Fred Jennie, was a departure from the time-tested Mauser. Heavy action and strength are well known attributes of the Weatherby Mark V, and the receiver is machined from block steel while the push feed design utilizes a large claw extractor feature inside the fluted bolt body and a plunger ejector in the face of the bolt. The rifle includes an adjustable single-stage trigger.

Perhaps the most recognizable feature of the Weatherby Mark V is its superlative claro walnut stock. Constructed in a 'California style', it exhibits a forward slope with a slender pistol grip at the bottom. A tapered forend is flattened at the bottom for a steady rest position. Originally produced in West Germany, the manufacturing was subsequently moved to Japan.

Variants include the familiar Mark V Deluxe, the Mark V VarmintMaster introduced in 1965 and the Crown Custom with such cosmetic touches as a damascus steel bolt and rosewood cap on the pistol grip. The Weatherby Mark V has often been embellished with gold, silver, carving and inlay. These presentation-grade rifles have been given to celebrities as a marketing tool.

The Savage Model-110 was introduced in 1958 as a competitor to the

WEATHERBY MARK V
COUNTRY OF ORIGIN:
United States
DATE
1957
CALIBRE
5.68mm (.224in)
WEIGHT
3.29kg (7.25lb)
OVERALL LENGTH
1117.6mm (44in)
FEED/MAGAZINE
Bolt action; 5-round internal magazine
RANGE
91.44m (100yds)

WEATHERBY MARK V DELUXE
COUNTRY OF ORIGIN
United States
DATE
1958
CALIBRE
6.53mm (.257in)
WEIGHT
3.86kg (8.5lb)
OVERALL LENGTH
1184mm (46.625in)
FEED/MAGAZINE
Bolt action; 5-round internal magazine
RANGE
91.44m (100yds)

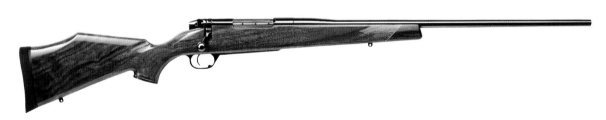

The 1958 Savage Model 110 rifle offered a competitive option to other bolt action sporting rifles.

popular Remington bolt-action rifles. Fed by either a detachable box or internal magazine with up to four rounds in a number of calibres, the Model 110 was the opposite of the over-engineered, more expensive rifles offered by the competition. It was inexpensive, and there was no doubt the user understood that. However, it proved accurate and affordable even with a barrel screwed to the receiver, a sloppy trigger and other parts that were cheap to produce. The rifle is still in production and includes Target and Varmint variants. The Model 110 has enabled Savage to continue to compete in the modern firearms market.

Sauer & Sohn of Germany produced the Model 90 bolt-action rifle with highly polished and checkered claro walnut stock and a barrel of cold forged Krupp steel with blued finish. Nine variants were produced before the model was discontinued in 2008, and the rifle is well known for its smooth action and lock.

Ruger

American firearms manufacturer Ruger surprised the firearms industry in 1966 with the single-shot bolt-action No.1 rifle. Although single shots had, many believed, become extinct in the way of new models, Ruger developed a rather economical method of manufacturing called investment casting and avoided costly machining in the production of the No.1 receiver. The No.1 was chambered in 30 different calibres and featured falling block action. It remains in production today.

SAVAGE M10-110T
COUNTRY OF ORIGIN
United States
DATE
1958
CALIBRE
7.62mm (.30in)
WEIGHT
3.18kg (7lb)
OVERALL LENGTH
1130mm (44.5in)
FEED/MAGAZINE
Bolt action; 4-round internal
or box magazine
RANGE
457.2m (500yds)

SAUER 90
COUNTRY OF ORIGIN
Germany
DATE
1972
CALIBRE
6.86mm (.27in)
WEIGHT
3.5kg (7.72lb)
OVERALL LENGTH
1067mm (42in)
FEED/MAGAZINE
Bolt action; 4-round internal
box magazine
RANGE
437.45m (400yds)

ABOVE: A hunter takes aim with a Ruger Model 1 rifle. The Model 1 debuted in the 1960s and revived interest in single-shot rifles at a time when many considered them obsolete.

Ruger followed the No.1 in 1968 with the Model 77, a rifle intended as a modern version of the famed Mauser 98. With its five-round integral box magazine, the rifle incorporates a few alterations from the original Mauser by designer L. James Sullivan, such as a plunger-style ejector rather than the original blade style.

Among other popular bolt-action rifles, the Accuracy International target and tactical offerings include the Varminter, known for its durability in inclement conditions. The Varminter offers the basic action and attributes of the company's AW rifle with the addition of a match grade barrel, eight-round staggered detachable box magazine and a coloured stock. An adjustable bipod and cheek piece are available. Another such rifle is the CZ 527 Varmint Kevlar, manufactured by the Czech firm of Ceská Zbrojovka Uherský Brod. Chambered for 5.66mm- (.223in-) calibre ammunition fed from a five-round detachable box magazine, the rifle exhibits a 'modified'

Mauser action. The Armscor M1700, originally conceived as a hunting rifle, has also found an application in 'varmint' control. Lightweight and chambered for the 4.5mm (.17in) HMR cartridge fed from a five-round internal magazine, it features a wooden stock and blued or stainless finish on metal components.

Tikka Series

Along with its popular Whitetail Hunter, recognized for value with a synthetic stock and adjustable trigger, Sako of Finland produces the Tikka series. One of these, the Tikka 3 includes two front locking lugs, a recessed face, plunger ejector and closed receiver machined from solid bar stock steel. The Tikka 3 stock is of quality walnut, and the barrel is free-floating and cold hammer forged. It is available in 7.82mm (.308in), 7.62x63mm (.30in) and other calibres. Another Sako offering is the innovative bolt action Quad, with four interchangeable free-floating barrels of four different calibres. Manufactured with a single stage trigger and with action based on the Sako P04R rifle, the Quad is fed from a detachable box magazine of up to 22 rounds. Its stock is either walnut or a glass fibre copolymer.

Austria's Steyr Mannlicher firm has been competitive since the 1990s with the Pro Hunter. Its stock is a molded high-density polymer, and the action is stainless steel with a pair of front locking lugs and two more at the rear. The barrel is stainless steel with a signature Mannlicher forged twist on the surface. The Robar Precision Hunter features an action similar to the Remington 700 series, a fluted stainless steel match grade barrel and

ACCURACY INTERNATIONAL VARMINT
COUNTRY OF ORIGIN
Great Britain
DATE
1980
CALIBRE
5.56mm (.22in)
WEIGHT
6kg (13.2lb)
OVERALL LENGTH
1155mm (45.47in)
FEED/MAGAZINE
Bolt action; 8-round detachable box magazine
RANGE
600m (656.17yds)

TIKKA T3 HUNTER
COUNTRY OF ORIGIN
Finland
DATE
1995
CALIBRE
7.82mm (.308in)
WEIGHT
3.1kg (6.83lb)
OVERALL LENGTH
1080mm (42.5in)
FEED/MAGAZINE
Bolt action; 3-round internal box magazine
RANGE
457.2m (500yds)

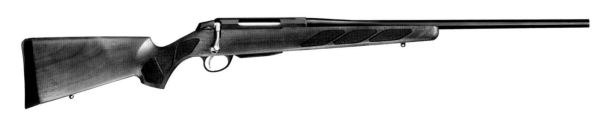

ABOVE: The light Tikka
3 rifle is produced in
several calibres by Sako of
Finland. It features four
interchangeable barrels
along with some plastic
components to reduce
overall weight.

a MacMillan hunting-style stock. Both the Pro Hunter and the Precision
Hunter are available in numerous calibres.

The modern design of the Browning Eurobolt includes a bolt turn of only
60 degrees, facilitating rapid loading, and a release lever placed atop the grip
for ease of use. The hog's back stock is favoured by stalking hunters. The
standard long bolt detachable magazine holds four 7mm (.275in) rounds,
while the magnum magazine holds three, and the rifle is chambered for 7mm,
.300 or .56mm ammunition.

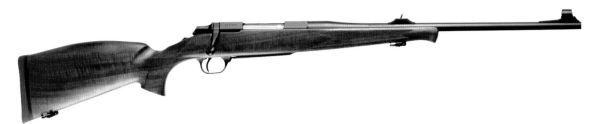

The Dakota Model 76, produced in Classic and Safari grades, reached the market in 1987 as a variant of the Remington Model 70 that incorporates Mauser 98 features as well. It includes a patented gas shield, bolt guide and bolt stop combination and was produced in several calibres, including 6.52mm (.257in) and 11.63mm (.458in).

The 5.58mm- (.22in-) calibre KK300 by Walther of Ulm, Germany, is a noteworthy single shot target rifle often manufactured with a thumbhole beechwood stock. Kriegskorte & Company of Stuttgart, Germany, has manufactured fine .22-calibre bolt-action target rifles since the 1950s. Known popularly as Krico, the company was originally founded to sell surplus Mauser rifles before embarking on the development of its own proprietary lines.

Semiautomatic Sporting Rifles

The Ruger semiautomatic blowback 10/22 rifle is one of the most popular firearms of its kind in history. Since 1964, more than five million have been manufactured. Tailored for hunting small game and target shooting, the 10/22 has been produced in a half dozen variants with a 10-round rotary or 25-round box magazine feeding 5.58mm- (.22in-) calibre ammunition. The simple construction of the 10/22 includes a barrel screwed to the frame. Numerous wooden and synthetic stocks are available and variants include the Carbine, Takedown, Sporter, Target, Tactical and Compact.

Browning's classic SA-22 was first produced in 1914 in the FN facility in Belgium. After 60 years of production in Belgium, manufacturing was

BROWNING EUROBOLT
COUNTRY OF ORIGIN
United States
DATE
1990
CALIBRE
7mm (.275in)
WEIGHT
3.18kg (7lb)
OVERALL LENGTH
1066.8mm (42in)
FEED/MAGAZINE
Bolt action; 4-round detachable magazine
RANGE
1000m (1093.6yds)

RUGER 10/22
COUNTRY OF ORIGIN
United States
DATE
1964
CALIBRE
5.7mm (.22in)
WEIGHT
2.3kg (5lb)
OVERALL LENGTH
940mm (37in)
FEED/MAGAZINE
Blowback; 25-round box magazine
RANGE
140m (153yds)

BROWNING SA-22

COUNTRY OF ORIGIN
Belgium
DATE
1914
CALIBRE
5.56mm (.223in)
WEIGHT
2.4kg (5.2lb)
OVERALL LENGTH
940mm (37in)
FEED/MAGAZINE
Semiautomatic; 16-round
integral tube magazine
RANGE
91.44m (100yds)

BAR HUNTING RIFLE

COUNTRY OF ORIGIN
Belgium
DATE
1972
CALIBRE
7.62mm (.30in)
WEIGHT
2.7kg (6lb)
OVERALL LENGTH
1000mm (41in)
FEED/MAGAZINE
Semiautomatic; 4-round
detachable box magazine
RANGE
550m (601.49yds)

transferred to Japan. The world's first production semiautomatic rifle, the SA-22, also known as the Browning 22 Semi-Auto Rifle, is fed by an integral tube magazine holding 11 long cartridges or 16 short. Highly prized by collectors, the SA-22 is often made with a walnut stock and blued steel finish on metal parts. Engraving is sometimes added since the rifle has a large face with the absence of a side ejector. The ejection of spent cartridges is downward to protect the user from debris and gasses.

In the 1970s, Browning began identifying its 7.62x63mm (.30in) semiautomatic hunting rifle as the BAR (BAR), which alluded to the storied history of the military rifle of the same name. Although the name is common, the two weapons share no other similarities. The BAR hunting rifle has been manufactured in four variants, Safari, LongTrac, ShortTrac and Lightweight Stalker. The Safari includes an engraved receiver and high quality stock along with the BOSS (Ballistic Optimizing Shooting System) that includes a muzzle brake to reduce recoil. The Lightweight Stalker features an aluminum alloy receiver without engraving and a synthetic stock.

The LongTrac and ShortTrac versions are based on the length of their bolt action and are manufactured with a plastic trigger assembly and walnut stock. These are available in the more upscale Stalker with a matte black finish and composite stock and the Mossy Oak with a composite stock and overall camouflage colouration.

The Remington Model 24 autoloading 5.58mm- (.22in-) calibre rifle debuted in 1922 and remained in production until 1935. Designed by the legendary John Browning, it is easily taken down with simple tools. Similar

in configuration to the Browning SA-22, the Model 24 was
succeeded by the Model 241 that differed slightly from its predecessor and
the SA-22 in the method of tightening the barrel. More than 107,000 of the
Model 241 were produced from 1935 to 1949.

Beginning in 1955, Remington produced the Model 740 semiautomatic
rifle that was eventually chambered for several calibres, including 5.58mm
(.22in) and 7.62x63mm (.30in). Its five-year production run ended in
1960, and more than 251,000 were manufactured. For the next 20 years,
the Remington Model 742 was tremendously popular with its rotating
breechblock, side ejection port and four-round internal magazine. Nearly
1.5 million were built.

With a checkered pistol grip and straight comb stock, the Remington
Model 7400 was in production from 1981 to 2004 chambered for numerous
cartridges and in three variants, the Carbine, Special Purpose and Synthetic.
In 2006, the Model 750 was introduced and it remains in production in two
major configurations, the Woodsmaster with improved gas operation, swivel
studs for a sling and stock and forend in highly polished walnut, and the
Synthetic, which incorporates a synthetic rather than walnut stock.

The first semiautomatic in the Remington 500 target series was the
550 Autoloader. The 5.58mm- (.22in-) calibre blowback rifle has a tubular
magazine that holds up to 22 short rounds.

Nylon 66

One of the more interesting Remington semiautomatic rifles is the
Nylon 66, which reached the market in late 1959 and featured a one-piece
stock of synthetic DuPont Zytel. The first mass production rifle to include
a stock made from a substance other than wood, the Nylon 66 was a risky
venture for Remington, and its leadership provided specs to DuPont,
which in turn developed the Zytel composite. The 5.58mm- (.22in-)
calibre Nylon 66 receiver was also constructed of Zytel, requiring little
maintenance and virtually no additional lubrication. Fed by a 14-round

**REMINGTON
MODEL 750**
COUNTRY OF ORIGIN
United States
DATE
2006
CALIBRE
7.62mm (.30in)
WEIGHT
3.4kg (7.5lb)
OVERALL LENGTH
990mm (39.1in)
FEED/MAGAZINE
Semiautomatic; 4-round
detachable box magazine
RANGE
550m (601.49yds)

CZ 511

COUNTRY OF ORIGIN
Czechoslovakia
DATE
1970
CALIBRE
5.56mm (.223in)
WEIGHT
(5.9lb)
OVERALL LENGTH
(39.3in)
FEED/MAGAZINE
Semiautomatic; 8-round
detachable box magazine
RANGE
91.44m (100yds)

tubular magazine, more than a million examples of the rifle were produced in five variants during interrupted production cycles through 1989.

From Ceská Zbrojovka Uherský Brod the CZ 511 semiautomatic .22-calibre rifle includes basic iron sights, an eight-round detachable magazine and a groove for mounting a scope. It is also outfitted with a walnut stock and swivel rings for mounting a sling.

Manufactured by Benelli of Italy, the R1 semiautomatic is a big game rifle chambered for 7.62x63mm (.30in), 7.62mm (.30in magnum) or 9.65mm

(.380in magnum) ammunition. Its simple yet strong design includes the ARGO – auto-regulating gas operating piston system – to substantially reduce recoil. The R1 is also favoured for its trigger. A synthetic Comfortech or AA grade walnut stock is available, and a Picatinny rail allows easy scope mounting. The rifle is fed by a three-round detachable box magazine.

A descendant of the Colt manufactured AR-15 semiautomatic civilian and M16 military assault rifles, the Colt Match Target Competition HBAR M6700 is available in semiautomatic mode only and readily

BENELLI R1
COUNTRY OF ORIGIN
Italy
DATE
1980
CALIBRE
7.62mm (.30in)
WEIGHT
3.31 (7.3lb)
OVERALL LENGTH
1168.4mm (46in)
FEED/MAGAZINE
Gas operated; 4-round detachable box magazine
RANGE
550m (601.49yds)

LEFT: Benelli of Italy was founded in 1967 and produces a variety of shotguns, rifles and pistols. A number of these are shown on display during a trade show held in Nuremberg, Germany.

COLT MATCH TARGET HBAR
COUNTRY OF ORIGIN
United States
DATE
1970
CALIBRE
5.56mm (.223in)
WEIGHT
3.86kg (8.5lb)
OVERALL LENGTH
1000m (39.5in)
FEED/MAGAZINE
Gas operated; 30-round detachable box magazine
RANGE
550m (601.49yds)

HK SR9
COUNTRY OF ORIGIN
Germany
DATE
1990
CALIBRE
7.62mm (.30in)
WEIGHT
4.95kg (10.9lb)
OVERALL LENGTH
1080mm (42.5in)
FEED/MAGAZINE
Semiautomatic; 5-round detachable box magazine
RANGE
550m (601.49yds)

adaptable from the target range to hunting. Chambered for the 14.1mm (.556in) NATO cartridge, it also accepts 5.56mm (.223in) Remington ammunition. It features a multi-lug rotary locking bolt, a raised cheek piece, pistol grip rest and often a 3x9 telescopic sight protected by rubber coating. Its synthetic grip and stock are finished in matte black.

Heckler and Koch Semiautomatic Rifles
One of the more popular Winchester semiautomatic rifles is the Model 77, introduced in 1955 and produced until 1963. Nearly 220,000 of the 5.58mm-(.22in-) calibre Model 77 have been sold, and the rifle includes a detachable box or tubular magazine under the barrel. The .22-calibre Model 74 included a tubular magazine loaded in the buttstock and was manufactured in two versions, the Sporting model and the Gallery Special.

Other noteworthy semiautomatic rifles include the Heckler & Koch Model 91, derived from the military G3 assault rifle. The Model 91 was specially crafted for the U.S. market and arrived in 1974. With a 20-round detachable box magazine, the rifle is chambered for the 7.82mm (.308in) Winchester cartridge and is a near clone of the G3 with the exception of a grenade launcher ring attached to the muzzle of the military G3. Although some states in the U.S. have tightened gun laws that prevent ownership of the Model 91, it has been successfully marketed to hunters.

Another Heckler & Koch offering, the SR9 sporting rifle includes an ergonomic thumbhole stock made of fibreglass reinforced by Kevlar. Its MSG90 buffer system dampens recoil substantially – perhaps more than any other semiautomatic rifle.

In 1972, firearms designer Mark Gwinn debuted his 5.66mm- (.223in-) calibre Bushmaster semiautomatic rifle incorporating improvements he believed were needed to the M16 assault rifle after his experience in Vietnam. Gwinn Firearms was later acquired by the Quality Products company and eventually became Bushmaster Firearms, located in Windham, Maine.

ABOVE: The Bushmaster XM15/AR15 series of rifles is popular with those who appreciate military-grade reliability in a hunting weapon.

Gwinn went on to develop a series of Bushmaster rifles that generally were intended for the hunting market and to improve on the related AR-15 semiautomatic civilian and M-16 military assault rifles. One of these was the 5.58mm- (.22in-) calibre Bushmaster Stainless Varmint Rifle with risers for the mounting of a scope, a synthetic stock and a highly polished aluminum barrel.

Lever-action Sporting Rifles

The name Winchester has been synonymous with lever-action rifles for more than a century, and the Model 94 is closely associated with the generic phrase 'deer rifle'. Originally introduced in 1894 and designed by John Browning, the Model 1894 (see chapter one) was substantially revised in 1964 in a modernization effort. Through the years, the rifle has most often been chambered for the famous .30-30 round, and more than 7.5 million of all versions have been manufactured.

The 1964 model is notable because several production changes proved controversial. Parts were switched from a process that machined them from solid block steel to a sintered process, forming them from metal powders. A stamped sheet metal cartridge lifter and pins that were hollow rolled rather than solid steel were perceived as lower quality despite the fact that performance was not adversely affected. Combined with the fact that the surface of the sintered steel did not accept bluing as well as the solid predecessor, these changes resulted in many users staying with pre-1964 Model 94 rifles. The idea is persistent today, and those Model 94 rifles made prior to 1964 command higher resale values.

In 1982, the cartridge ejection system of the Model 94 was changed from the top to an angled configuration to mount telescopic sights and maintain sales momentum against stiff competition. Precision-machined CNC parts and solid pins were reintroduced in 1992. A controversial safety was reengineered in 2003.

WINCHESTER 94 - LEGACY 22
COUNTRY OF ORIGIN
United States
DATE
1972
CALIBRE
5.58mm (.22in)
WEIGHT
2.72kg (6lb)
OVERALL LENGTH
994mm (39.125in)
FEED/MAGAZINE
Lever action; 21-round internal magazine
RANGE
91.44m (100yds)

Other noteworthy Winchester lever action rifles include the Model 71, Model 88 and the Model 9422. The Model 71 was introduced in 1935 as a replacement for the Model 1886 and Model 1895 rifles and essentially was an updated Model 1886 chambered for the 8.83mm-(.348in-) calibre Remington cartridge. It was the only Winchester rifle chambered for that round and developed a loyal following. However, high manufacturing costs and less expensive competitors resulted in its discontinuation despite a production run that lasted until 1958 with about 47,000 completed.

The Model 88 was issued in 1955 and marked the 100th anniversary of the company. The rifle was produced in two variants, the full size rifle with a 559mm (22in) barrel and the carbine with a 483mm (19in) barrel. Both were chambered for 5.58mm- (.22in-) calibre ammunition, and they were fed from either a tubular magazine under the barrel or a detachable box magazine.

The Model 88, sometimes referred to as the Centennial Model, may be considered a hybrid because of its use of a lever and a rotating bolt with three lugs. This combination allowed the use of some short case cartridges with spritzer bullets. Despite the fact that more than 280,000 were produced by 1968 and the Model 88 sold more than any other Winchesters except the Model 1892 and Model 1894, production was discontinued by the early 1970s.

The 5.58mm- (.22in-) Model 9422 reached the market in 1972 and presented the image of the traditional lever action rifle; however, it was produced with a groove for the mounting of a scope. Those who enjoyed the history of the Wild West gravitated toward the rifle due to its frontier look and reliability. The rifle was discontinued in 2005. Chambered for several different .22-calibre rounds, its tubular magazine holds up to 21 cartridges.

Many lever-action rifle aficionados will agree that the greatest rival to the Winchester 94 is the Marlin Model 336. Introduced in 1948 and in continuous production since then, the Model 336 can trace its lineage to the Marlin Model 1893, which incorporated an innovative locking system and two-piece firing pin. The Model 1893, also with a solid top receiver and side ejector, was produced until 1936 and was actually replaced in 1937 by the

WINCHESTER MODEL 88
COUNTRY OF ORIGIN
United States
DATE
1955
CALIBRE
7.82mm (.308in)
WEIGHT
3.29kg (7.25lb)
OVERALL LENGTH
1079.5mm (42.5in)
FEED/MAGAZINE
Lever action; 4-round detachable box magazine
RANGE
91.44m (100yds)

MARLIN 336 DELUXE

COUNTRY OF ORIGIN
United States
DATE
1948
CALIBRE
7.62mm (.30in)
WEIGHT
3.18kg (7lb)
OVERALL LENGTH
1080mm (42.25in)
FEED/MAGAZINE
Lever action; 4-round internal tube magazine
RANGE
274.32m (300yds)

MARLIN 336Y

COUNTRY OF ORIGIN
United States
DATE
2011
CALIBRE
7.62mm (.30in)
WEIGHT
2.95kg (6.5lb)
OVERALL LENGTH
844.55mm (33.25in)
FEED/MAGAZINE
Lever action; 4-round internal tube magazine
RANGE
274.32m (300yds)

Model 1936. In truth, the Model 1936, soon simply renamed the Model 36, was only slightly different than the old Model 1893, with a modified stock, sights and forend.

The introduction of the Model 336 brought the ideas of Marlin designer Thomas R. Robinson, Jr., forward. The Model 336 included coiled springs rather than the flat springs used in earlier Marlin rifles, an open ejector port in the side of the receiver, an improved extractor and cartridge carrier, a rounded and chromed breech bolt and a flat receiver top that facilitated the mounting of a scope. The majority of the rifles produced featured a fine walnut stock and a tubular magazine. Chambered for several different rounds, including the Winchester 7.62x51mm (.30in), the 8.89mm (.35in) Remington and the 7.82mm (.308in) Marlin Express, the Model 336 is also noted for its ease of dismantling and cleaning, its simple construction and its accuracy.

A carbine version of the Marlin 336 has been produced along with Marauder and Trapper variants and the 336Y, a rifle for younger users. Marlin also produced a less expensive Model 336 under the Glenwood brand. Among other cost-saving measures the stock of the Glenwood version was beechwood rather than walnut.

Another Marlin lever-action favourite is the Model 39A, which has essentially been in continuous production since 1891 when its parent, the Model 1891, was introduced as the first lever-action rifle chambered for the 5.58mm- (.22in-) calibre long rifle cartridge. After a series of revisions, the Model 39A appeared in 1937. The rifle differs little from its other predecessors,

including the Model 1892, Model 1897 and Model 39. It was followed by the Golden Model 39A, with a distinguishing gold trigger, in the 1950s.

With a solid top receiver and side ejector, the Golden Model 39A readily accepts a scope. Its hardware is machined from forged steel and the stock is walnut with a pistol grip. The tubular magazine holds up to 26 5.58mm (.22in) short rounds or up to 21 long rifle cartridges. One criticism of the Golden Model 39A and its close relatives is its takedown, requiring the turn of only one screw to separate the rifle into two pieces.

The Savage Model 99 emanated from the company's Model 1895, the first production hammerless lever action rifle. Soon after production began, the Model 1895 was modified slightly and renamed the Model 1899 – later shortened to Model 99. These closely related rifles utilized a rotary magazine that held five cartridges in a spool for many years before it was replaced by a detachable box magazine. The rotating magazine allowed the use of spritzer bullets, and the rear locking and cam combination facilitated the use of heavier ammunition as well. Through the years, the Model 99 was chambered for numerous cartridges, and the rifle was produced until 1998.

The labour-intensive production of the Savage Model 99 required a number of hand-performed steps; therefore, its cost was high compared to other rifles. Nevertheless, its fine tolerances and outstanding performance kept it in production for more than a century.

The Browning BL-22 offers the look of the Wild West to both experienced and novice operators. One of this 5.58mm- (.22in-) calibre rifle's best

MARLIN 39A
COUNTRY OF ORIGIN
United States
DATE
1891
CALIBRE
5.58mm (.22in)
WEIGHT
2.7kg (6lb)
OVERALL LENGTH
97mm (40in)
FEED/MAGAZINE
Lever action; 19–26-round tube magazine
RANGE
91.44m (100yds)

SAVAGE MODEL 99
COUNTRY OF ORIGIN
United States
DATE
1899
CALIBRE
7.7mm (.303in)
WEIGHT
3.18 (7lb)
OVERALL LENGTH
1066.8mm (42in)
FEED/MAGAZINE
Lever action; 5-round detachable box magazine
RANGE
274.32m (300yds)

BROWNING BL-22

COUNTRY OF ORIGIN
United States
DATE
1969
CALIBRE
5.58mm (.22in)
WEIGHT
2.27kg (5lb)
OVERALL LENGTH
933.45mm (36.75in)
FEED/MAGAZINE
Lever action; 15-round
internal tube magazine
RANGE
91.44m (100yds)

attributes is its 33-degree lever action that ejects spent cartridges from the side and chambers the next round with ease, and the trigger travels with the lever to prevent pinching an entry level user's finger.

Produced in six grades, the BL-22 offers an attractive walnut stock and is complemented by a blued barrel and receiver finish with the receiver grooved to easily mount a scope. Adjustable sights offer a high degree of accuracy even for beginners, and the straight grip enhances control. While the BL-22 accommodates long and short 5.58mm- (.22in-) calibre rounds, it will hold up to 15 .22-calibre long rifle cartridges.

The Henry Repeating Arms Company of Bayonne, New Jersey, has a rich history that stretches back more than 150 years. One recent addition to its family of lever action rifles is the Big Boy, introduced in 2001 and chambered for 11.17mm- (.44in-), 11.43mm- (.45in-) and 9.06mm- (.357in-) calibre ammunition. The original Big Boy was the .44. It was followed soon by the .45, and the .357 entered the market in 2006. The rifle features an octagonal

barrel, 10-round tubular magazine and receiver machined from solid brass.

Complementing the sleek .22-calibre Henry Golden Boy lever action series, the Big Boy is the first .44-calibre Henry repeating rifle made in the United States with a solid brass receiver since the original Henry of 1860. While the .45 and .357 versions are reasonable when it comes to recoil, the .44 exhibits a considerable kick.

The action of the Big Boy is quite similar to that of the Marlin Model 336 with its solid receiver and round bolt with a single rear-positioned locking lug. The Model 336, however, loads through a port in the receiver while the Big Boy loads through a port in the forward end of the tube magazine. The Big Boy trigger action is a strong attribute, and the straight grip stock is fine walnut.

Other Modern Sporting Rifles

Several slide-action sporting rifles have earned their places among the finest of their genre, and the venerable Remington Model 12 is one of the top selling sporting rifles in history with more than 830,000 produced from 1909 to 1936. Its tubular magazine, situated under the barrel, holds up to 22 rounds of 5.58mm- (.22in-) calibre ammunition, and its resemblance to Remington slide-action shotguns is unmistakable.

Contemporary with the Model 12, Remington generated the Model 14. Prior to 1910, company designer John Pederson was charged with developing a rifle to compete with the lever-action Winchester Model 1894. The slide-action result included a bolt that unlocked by pressing a button through the ejection port, a spiral magazine that prevented rounds from touching one another nose to tail and loading through an opening in the magazine between the forward end of the rifle and the receiver.

The Model 14 may be considered an upgraded Model 12 capable of firing more powerful rounds. Introduced in 1912, the original rifle was produced until 1934 along with a carbine variant with a shortened barrel. The Model 14½ was virtually identical but chambered for Winchester .38-40

HENRY BIGBOY
COUNTRY OF ORIGIN
United States
DATE
2001
CALIBRE
11.17mm (.44in)
WEIGHT
3.95kg (8.7lb)
OVERALL LENGTH
978mm (38.5in)
FEED/MAGAZINE
Lever action; 10-round internal tube magazine
RANGE
91.44m (100yds)

LEFT: Since its debut in 2001, the Henry Big Boy lever action rifle has become a favourite of shooters who prefer a nostalgic look and a large calibre round.

REMINGTON MODEL 14
COUNTRY OF ORIGIN
United States
DATE
1912
CALIBRE
5.58mm (.22in)
WEIGHT
3.52kg (7.75lb)
OVERALL LENGTH
1086mm (42.75in)
FEED/MAGAZINE
Slide action; 5-round tubular magazine
RANGE
91.44m (100yds)

WINCHESTER MODEL 61
COUNTRY OF ORIGIN
United States
DATE
1932
CALIBRE
5.58mm (.22in)
WEIGHT
2.7kg (5.9lb)
OVERALL LENGTH
1047.75mm (41.25in)
FEED/MAGAZINE
Slide action; 10-round tubular magazine
RANGE
91.44m (100yds)

and .44-40 ammunition. It was produced from 1914 to 1931 and was also made in a carbine version. The Model 141 upgraded the original Model 14 with improved adjustable sights. It was made in a carbine variant from 1935 to 1950.

In 1936, Remington marketed the Model 121, a slide-action companion to the Model 12 and Model 14. It featured a larger stock with a pronounced pistol grip and a checkered steel buttplate. Its tubular magazine held up to 20 rounds of 5.58mm- (.22in-) calibre ammunition.

In 1922, Browning introduced the slide-action Trombone Rifle, which was produced at the FN facility in Belgium until 1974. The Trombone Rifle was named due to its slide action and proved to be a remarkably popular .22-calibre firearm with a tubular magazine and hammerless action. More than 150,000 were manufactured during its half-century production run.

The Winchester Model 61 slide-action rifle was introduced in 1932 as the fourth in a series of such weapons that stretched back to 1890. The Model 61 was manufactured in several versions that were chambered individually for .22in-calibre long, .22in-calibre long rifle or .22in-calibre short ammunition, or for all three, and finally for only the .22in Winchester magnum cartridge. A takedown rifle that was simple to disassemble, the Model 61 was hammerless with either a round or octagonal barrel. It included a tubular magazine and maple stock, and more than 342,000 were sold during production that ended in 1963.

A second Winchester slide-action rifle of the period was the Model 62 with an exposed hammer that was also meant to replace earlier Winchester efforts to popularize the style.

Sporting Rifle Odds and Ends

A few modern sporting rifles are unusual in some respect, and one of these is the richly appointed Westley Richards Droplock Double Rifle. When the British company introduced its fine firearms around the turn of the twentieth century, it set a standard of elegance that is simply stunning. These expensive rifles are famous for their hand-detachable lock action with seven components and for their lever work. They are highly embellished with engraving and fitted with a traditional recoil pad, gold oval and top quality walnut presentation wood stock with checkering. The rifles are hand crafted with approximately 600 hours invested in each one.

For black powder enthusiasts, Remington began producing the 12.7mm- (.50in-) calibre Model 700 Muzzleloader in 1966. A steel plug and percussion nipple sit along the breech of the rifle that is otherwise based on the Model 700 bolt-action series. A cylinder shaped hammer sits inside the breechblock rather than the firing pin of the bolt-action Model 700. Their synthetic stocks come in several colours.

In 1985, well-known gunsmith Tony Knight introduced his Knight MK-85 muzzleloader. The Knight rifle features the percussion nipple positioned at the rear of the barrel behind the powder charge rather than beside it. The result is significantly greater reliability when firing the 12.7mm- (.50in-) calibre rifle. Knight is credited, through his inventiveness, with helping to launch the great resurgence of interest in blackpowder rifles some 30 years ago. The MK-85 stock is typically black synthetic.

BELOW: The British manufacturer Westley Richards is renowned for its highly embellished and well appointed rifles and shotguns, while developing innovative production patents and fabricating firearms for the royal family.

By Appointment   H.M. The King

WESTLEY RICHARDS
HIGH GRADE GUNS & RIFLES
New 1926 Model—Wide Opening Gun and Improved Ejection

The Celebrated ·318-bore Accelerated Express Rifle

" The American Rifleman " says :
" The prestige for *all-round* quality in Kenya is now held by the Westley Richards ·318-bore."

NOTE TO SHOOTERS
See that the Rifle bears the " Westley Richards " name, which is the guarantee of superiority.

23, Conduit Street, London, W.1

Glossary

Action The working mechanism of a firearm, responsible for the main activities of loading, firing and ejecting. Action is also used sometimes in the same way as 'receiver'.

Bolt The part of a firearm which usually contains the firing pin or striker and which closes the breech ready for firing.

Blowback Operating system in which the bolt is not locked to the breech, thus it is consequently pushed back by breech pressure on firing and cycles the gun.

Bore The interior section of a gun's barrel.

Boxlock A type of action in a break-open gun where all of the lockwork is contained within a box-like housing. Boxlocks are the most common type of double-barrelled shotgun mechanism, being relatively inexpensive to manufacture and extremely robust.

Breech The rear of the gun barrel.

Breech-block Another method of closing the breech which generally involves a substantial rectangular block rather than a cylindrical bolt.

Bullpup Term for when the receiver of a gun is actually set in the butt behind the trigger group, thus allowing for a full length barrel.

Carbine A shortened rifle for specific assault roles.

Centrefire A cartridge that has the percussion cap located directly in the centre of the cartridge base.

Chamber The section at the end of the barrel which receives and seats the cartridge ready for firing.

Closed bolt A mechanical system in which the bolt is closed up to the cartridge before the trigger is pulled. This allows greater stability through reducing the forward motion of parts on firing.

Compensator A muzzle attachment which controls the direction of gas expanding from the weapon and thus helps to resist muzzle climb or swing during automatic fire.

Delayed blowback A delay mechanically imposed on a blowback system to allow pressures in the breech to drop to safe levels before breech opening.

Double action Relates to pistols which can be fired both by cocking the hammer and then pulling the trigger, and by a single long pull on the trigger which performs both cocking and firing actions.

Ejector A system for throwing the spent cartridge cases from a gun.

Extractor A system for lifting spent cartridge cases out of the chambers, making them easily removed by hand.

Flechette An bolt-like projectile which is smaller than the gun's calibre and requires a sabot to fit it to the barrel. Flechette rounds achieve very high velocities.

Gas operation Operating system in which a gun is cycled by gas being bled off from the barrel and used against a piston or the bolt to drive the bolt backwards and cycle the gun for the next round.

GPMG Abbreviation for General Purpose Machine Gun. A versatile light machine gun intended to perform a range of different roles.

Guage The calibre of a shotgun bore. The term relates to the number of lead balls the same diameter as the bore that it takes to make 1lb (0.45kg) in weight.

LMG Abbreviation for Light Machine Gun.

Locking Describes the various methods by which the bolt or breech block is locked behind the chamber ready for firing.

Long recoil A method of recoil operation in which the barrel and bolt recoil for a length greater than that of the entire cartridge, during which extraction and loading are performed.

Muzzle brake A muzzle attachment which diverts muzzle blast sideways and thus reduces overall recoil.

Open bolt A mechanical system in which the bolt is kept at a distance from the cartridge before the trigger is pulled. This allows for better cooling of the weapon between shots.

PDW Abbreviation for Personal Defence Weapon. A compact firearm, smaller than a regular assault rifle but more powerful than a pistol, intended as a defensive weapon for personnel whose duties do not normally include small arms combat.

Receiver The body of the weapon which contains the gun's main operating parts.

Recoil The rearward force generated by the explosive power of a projectile being fired.

Recoil operated Operating system in which the gun is cycled by the recoil-propelled force of both barrel and bolt when the weapon is fired. Both components recoil together for a certain distance before the barrel stops and the bolt continues backwards to perform reloading and rechambering.

SAW Abbreviation for Squad Automatic Weapon.

Self-loading Operating system in which one pull of the trigger allows the gun to fires and reload in a single action.

Shaped charge An anti-armour charge designed to concentrate the effect of an explosive warhead by focusing a cone of superheated gas on a critical point on the target.

Short recoil A compressed version of recoil operation in which the barrel and bolt move back less than the length of the cartridge before the bolt detaches and continues backwards to perform reloading and rechambering.

Index